MUDSLINGERS

MUDSLINGERS

A TRUE STORY OF AERIAL FIREFIGHTING

AN AMERICAN ORIGINS STORY

TIM SHEEHY

PERMUTED PRESS

A PERMUTED PRESS BOOK
ISBN: 979-8-88845-205-9
ISBN (eBook): 979-8-88845-206-6

Mudslingers:
A True Story of Aerial Firefighting (An American Origins Story)
© 2023 by Tim Sheehy
All Rights Reserved

Cover design by K Mita @ Sparkle Bomb Studio
Cover photo by Austin Rauh

This is a work of nonfiction. All people, locations, events, and situations are portrayed to the best of the author's memory.

No part of this book may be reproduced, stored in a retrieval system, or transmitted by any means without the written permission of the author and publisher.

PERMUTED PRESS

Permuted Press, LLC
New York • Nashville
permutedpress.com

Published in the United States of America
1 2 3 4 5 6 7 8 9 10

CONTENTS

American Origins Stories ix
Chapter 1 - Prologue 1
Chapter 2 - American Mudslingers 17
Chapter 3 - Talk-Talk 46
Chapter 4 - The Road Less Traveled 77
Chapter 5 - The First Fire Bombers 96
Chapter 6 - Changing Lanes 117
Chapter 7 - Fighting Fire 139
Chapter 8 - Let it Burn 156
Chapter 9 - Flyboys 184
Chapter 10 - The Super Scooper 203
Chapter 11 - Accidents Will Happen 228
Chapter 12 - Airtankers 250
Chapter 13 - The Industrial Revolution 268
Chapter 14 - Withdrawal 282
Epilogue 301
Glossary 311
Selected Bibliography & Sources 319
Acknowledgments 325

To the brave men and women who have perished fighting fire from the air, and to their families who carry on their legacy. You are a small and elite group who deserve our eternal gratitude.

All author proceeds from this book go to benefit the Montana Firefighter Fund and the Wildland Firefighter Foundation, charitable organizations established to benefit all wildland firefighters, air and ground (and their families) who are injured or have perished in the line of duty. Proceeds also support the United Aerial Firefighters Association, the only trade organization focused exclusively on the safety and operational excellence of the aerial firefighting industry.

AMERICAN ORIGINS STORIES

Recorded history lays out a fascinating trajectory for mankind. But much of that history is devoted to familiar topics. In the case of the United States, historical narratives tend to focus on events and people of enormous and obvious importance: military conflicts from the Revolutionary War to World War II; biographies of former presidents or industrial leaders or artists; accounts of key historical events, such as the civil rights movement or the first lunar landing or the birth of the internet. These are what I like to call the mountains and oceans of history, the massive events that dictate the flow of the rivers, the weather patterns that color our lives, the location and complexion of human habitation…and, in many ways, the course of human history. These events and people are usually well-researched and recorded through dozens of magnificent books, documentaries, films, theatrical presentations, interviews, and stories. They are History—with a capital "H."

But between these mountains, oceans, and rivers of history are often fascinating streams and meadows that are overlooked and unrecorded (or at least under-recorded).

These are the lesser-known events and trends that are hidden away in the memories of everyday people who haven't gotten the chance to tell their stories. Where do these end up in recorded history? The truth is, they usually don't. They fade with the passing of those who might have shared the stories. And, inevitably, the texture of those events becomes lost in the sands of time. Sometimes the stories are overlooked because they weren't pivotal enough to be remembered; sometimes they don't fit a narrative shaped by academia or popular culture. In short, they are deemed "irrelevant." And yet, while some of these events might seem small when compared to the broader arc of history, that doesn't mean they are insignificant, and it doesn't mean they aren't interesting in their own right. Buried in that pile of forgotten tales are stories that have deceived us about their significance. They are small and forgettable in size, yet their overall impact on history can be felt for generations. Stories like the British Codebreakers for Project ULTRA at Bletchley Park in World War II. A relatively small team placed within a massive historical upheaval that ended up being at a fulcrum of history. These are events—and people—that punch above their weight class. I like to refer to them as the islands of history.

This book, *Mudslingers*, is the first installment in a series called *American Origins Stories*. These are books exploring the tales of islands of American history. Fascinating subject matter that I have had personal experience with and that, I feel, hasn't been shared in a condensed and digestible fashion. Since these stories are told from a first-person perspective, there will inevitably be some off-roading into personal storytelling that is not always totally relevant to the subject matter. Furthermore, being that they are not strictly history books, there will be people, places, and things I don't mention that I probably should have mentioned; it's a difficult balancing act, and I'm sure I failed in some spots. But I have done my best to shed a light on subjects that I think are worthwhile.

I hope you learn something in this series, and I hope you enjoy the stories.

CHAPTER I

PROLOGUE

JULY 19, 2020
MINDEN, NV

We have been languishing in the desert sun for weeks. Waiting, sleeping, reading, playing children's games to fend off boredom. With the mid-summer heat baking the tarmac and everything on it like a fried egg, sitting outside in the American desert has been exhausting. We sit next to our thirty-million-dollar aircraft, checking it over and over again each day, making sure it's ready in case we are needed. It almost feels like we are the crew from *Flight of the Phoenix*, stranded in the middle of the desert, awaiting rescue. Except we aren't. We are aerial firefighters, awaiting the order to launch an attack on a fire with our Super Scooper. We've been forced to sit outside all day, every day, because COVID-19 rules preclude entry into the government facility where we are based.

Here's the thing most people don't realize about fighting wildfires. As with urban firefighting and military combat deployments—and a handful of other professions seemingly (and not unjustly) defined by danger—the job generally consists of the following: prolonged periods of boredom and restlessness punctuated by bursts of almost unfathomable intensity or violence.

With each passing year, the concept of a "season" seems less applicable to the climatological phenomenon of wildfires. What once was a reasonably well-defined season following a predictable geographic pattern—the Southeast in winter and early spring, the Southwest in Spring, the Midwest in early summer, Big Sky Country in late summer, the Southwest and California in the fall—has morphed into what feels like a never-ending sprawl of heat and flame and smoke, stretching not just across the arid and timber-heavy regions of the United States, but all over the globe. Blame climate change, urban–wildland interface, poor forest management, or all of the above—the simple fact is that wildfires are getting worse. For those on the front lines of the ceaseless wildfire battle, the reasons behind the escalation are less important than the fact of its existence and the likelihood that things are going to get much worse before they get any better.

But this much is also true: wildfire management is besieged by territoriality, political bickering and partisanship, and a plethora of overlapping governmental agencies and private commercial ventures, all combining to form a splintered infrastructure that leads to delayed responses and a generally inefficient approach to the job that is not only frustrating but also counterproductive.

Two flight crews—a total of five pilots (two captains, two first officers, and one alternate) and four mechanics—from my company, Bridger Aerospace, have been dispatched from our headquarters in Bozeman, Montana, to a firebase here in the Eastern Sierras, roughly fifty miles south of Reno, Nevada, and thirty miles east of Lake Tahoe. We are based here under the direction of the state of Nevada, with whom we are contracted to provide aerial support as needed for fires that break out in the region.

As needed.

That's the complicated part. You would think that "need" is easily quantified. If there is a fire nearby and we have planes at the ready, then we scramble and take off and fly as quickly as possible to the fire zone, where we would then become participants in a carefully choreographed, but undeniably risky, ballet of planes and helicopters large and small, some providing oversight and guidance—*eyes in the sky*—and others dropping water and chemical flame retardant in an attempt to squelch or at least mitigate the fire so that ground crews can do the hard, up-close work of containment.

Indeed, sometimes this is exactly the way it works. Other times, you sit and wait for a call that never comes, even though you are close enough to assist, because some other agency has already dispatched a different company or because you've been told to wait for an opportunity elsewhere, which is how a company based in Montana might sit in California or Nevada and eventually be summoned to Arizona, while another company from Kentucky or Florida gets called to Nevada. Or perhaps you are simply too far back in the pecking order and waiting for a fire of sufficient magnitude to require your services because, frankly, sometimes the best strategy is to simply let the fire burn.

It is a bureaucratic pretzel that can confound the most experienced beltway rangers, and for a western neophyte like me, it has been utterly frustrating. Fires are burning, people need help, and here we sit. Moral and ethical frustrations aside, there are also very real commercial considerations. I have bet my entire financial life (and in one instance, my actual life) on these aircraft and this company. The government bean counters have no clue of the herculean effort—financially and personally—required to build an operation like this. But because of some clause in a contract that no one really understands, we are grounded.

Since being summoned here and placed on a state contract, we have spent two weeks passing time in hotel rooms or in the back of an airport trailer, playing cards, reading, fine-tuning our equipment over and over and over. There have been opportunities to fly—hot spots that probably could have used our help—but we have remained on standby, waiting for the moment when we will finally be able to utilize one of the two impressive, and impressively expensive ($30 million apiece), new aircraft in our fleet. Six years ago, my pregnant wife, infant daughter, and I were living in a tent on a plot of land that we had bought near Bozeman, our plans no more ambitious than to establish a home in a part of the country we found breathtakingly beautiful. There were vague intentions to build a ranch and fund the operation with the purchase of a twin-engine plane that I would operate and manage as a fledgling business devoted mostly to surveillance—assisting public safety agencies with search-and-rescue missions or helping ranchers locate lost livestock, for example.

I was a U.S. Naval Academy graduate and a former Navy SEAL officer who served multiple deployments in Iraq and Afghanistan. I had been wounded in action, and with a medical hold on my mission status and the wars winding down, my wife and I decided we had done our part. Looking beyond the Navy, I wasn't interested in the chase that motivated so many of my classmates—smart, ambitious men and women who took jobs in the private sector, typically in consulting or finance, where high six-figure paychecks awaited. We chose a lifestyle first, one rooted in the beauty of nature and tied to my life-long love affair with flying, but also committed to a mission we believed in—public service—and then hoped we could figure out a way to make it viable. Imagine my astonishment when viability gave way to prosperity, with one plane becoming two, and then four…then eight. By the time we took up temporary residence at Minden-Tahoe Airport, Bridger Aerospace had grown to a fleet of sixteen aircraft and nearly one hundred employees, making it one of the largest providers of aerial firefighting, maintenance, and technology services in the industry. Including our sister company at the time, Ascent Vision Technologies (AVT), we employed more than two hundred engineers, pilots, maintainers, and support staff providing critical technology development for the allied warfighter and aircraft for the American firefighter.

Simply put, we were successful beyond what I ever envisaged for my little enterprise. I was, by necessity, a businessman now, every bit as focused on ledgers and finance and spreadsheets and investment strategies as the Naval Academy classmates from whom I had diverged. But I was something else as well. I was a pilot, trained to work and fly alongside my employees—a talented and hard-nosed collection of former bush pilots and veterans who are less like the cowboys you might expect and more like the steady, cold-blooded professionals you find in the seat of an F-14. Cowboys simply don't last long in today's version of aerial firefighting—they die or get fired, victims of their own recklessness.

They are the best, and I consider myself blessed to be not just their boss, but their colleague. And make no mistake—when we are in the air, slicing through flames or getting buffeted by updrafts and fire-driven turbulence, that is exactly my role: teammate, crew member. These guys all have more experience fighting fires than I do, sometimes decades more, and once we are in the plane, I defer to their knowledge and experience. We are a team, committed to a mission of service that provides a feeling of unity unlike anything I have known outside a military deployment.

In the interest of transparency, it should be said that there is no ambiguity in fighting fires, no questioning of purpose or reason. We know what we are doing and why we are doing it. We know that each trip into the air provides relief and assistance to someone on the ground: homeowners in danger of losing their property, ranchers whose livestock is threatened, ground crews working in potentially lethal conditions. To them, we are the air cavalry, the good guys providing help, and there is a simplicity and a clearness of purpose that makes the job enormously appealing.

I am proud of my military service, proud of the men and women with whom I served (my wife is a Marine Corps veteran and a fellow graduate of the Naval Academy). But there were times I wondered, and I wonder even now, as the Forever War finally winds down, exactly what we accomplished in Afghanistan. It was easy to see our progress on the ground each day, during specific missions, but as we look back at the sacrifice years after our disastrous retreat, it's much harder. Here, in the smoky skies over the Sierras, everything is incongruously clear. It is a mission at once devastatingly complex and dangerous, yet gloriously simple.

Contain the fire. Extinguish the fire. Save lives. Save property. And make it home!

★ TIM SHEEHY ★

The waiting is the hardest part, and we've been doing a lot of that in recent weeks. We have a maintenance crew comprised of the best in the business, but there are only so many times you can clean and check an aircraft as it sits idly on the tarmac, waiting for a chance to fulfill its purpose—especially when that aircraft is brand spanking new. We pass the time checking the internet for weather forecasts that could indicate not just a threat of ignition, but of rapid evolvement.

Hot…dry…windy.

But even the confirmation of fire is no guarantee that we will be dispatched. There are five different federal agencies responsible for wildland fire management: The United States Department of Agriculture's Forest Service, the Department of the Interior's Bureau of Indian Affairs, the Bureau of Land Management, the U.S. Fish and Wildlife Service, and the National Park Service. All or some of which, at any given time, will take precedence over various state agencies, such as the Nevada Division of Forestry, which has contracted us for this assignment. The bureaucratic bullshit—who gets sent where and which aircraft are deemed most appropriate for the battle—is frequently hip-deep, and we are in the thick of it now.

On this assignment, there is frustration on multiple levels. We want to be in the fight…somewhere. We want to help. But there is an undeniable economic component to the anxiety, as our contract with the state of Nevada is less flexible and generous than contracts with some of the federal agencies for the simple reason that state coffers tend to be less deep than their federal counterparts. On a federal contract, you are generally paid by the day and the hour, whether your planes are in the air or on the ground. Not so with this contract. Being the first year of COVID and with government dysfunction across many agencies, our federal contract never came through in time for the season. So we are flying for the state of Nevada on a contract we had written with Ron Bollier, the Fire Management Officer for Nevada. Time on the ground is time uncompensated, and we've been burning a hole in our pockets for two weeks now—a hole made more noticeable by the presence of the bright yellow-and-red Super Scooper languishing on the runway. Ron did our company a great favor by helping us get in the door, and we did him a great favor by providing him a premium grade asset at rock bottom prices.

Essentially a magnificent flying boat whose sole purpose is to fight wildfires, the Super Scooper is a Canadair CL-415EAF (Enhanced Aerial Firefighter) manufactured by Viking Air (Now De Havilland). But it takes only one glimpse of the aircraft in action to understand why "Super Scooper" is exactly the right moniker. The CL-415EAF is an imposing beast, standing thirty feet tall with a wingspan of roughly one hundred feet—nearly as wide as a Boeing 737. It weighs 30,000 pounds when empty, which is to say, before it fulfills its appointed duty of loading up to 1,400 gallons of water in its brilliantly configured belly. Fully loaded, prior to drop, the CL-415EAF weighs 41,000 pounds.

Ungainly appearance and heft notwithstanding, the Super Scooper is an impressively agile and versatile aircraft, as it must be to fulfill its job requirement, which basically consists of flying to a body of water (one that is hopefully not far from the fire zone), then skimming the surface of the water and using a pair of probes to "scoop" water into its payload (while never coming to a stop) before flying off to an active fire and dumping its entire load on a designated spot, typically from an altitude of less than one hundred feet. Essentially, the Super Scooper tosses a giant pail of water onto a fire—which, interestingly, was precisely the strategy once employed by aerial firefighters of a different era. And the pilots who engage in these maneuvers must navigate in and around windswept canyons and mountains, darting in and out of cloudbanks thickened by black smoke and flame, their aircraft buffeted by unpredictable turbulence, all the while coordinating with other aircraft in the vicinity.

In commercial passenger aviation, if one plane comes within a *mile* of another plane, it is considered an egregious error and must be recorded and deconstructed in an incident report. In aerial firefighting, there might be more than a dozen aircraft above an area of less than a thousand acres—air attack planes supervising and providing guidance from the highest vantage point; scoopers, tankers, and helicopters dropping water and chemical retardant closer to the ground; drones or mapping aircraft high overhead—all circling and waiting, relying on a combination of radio communication, visual acuity, muscle memory, and intuitiveness. The fact that crashes today are rare—though certainly not unheard of—is something of a miracle. Indeed, the skill required of aerial firefighters is no less demanding, and the job no less dangerous, than that of a combat pilot.

The latest model of Super Scooper is one of the most recent innovations in an industry that has been slow to change and beset by administrative, jurisdictional, and logistical challenges. Fighting wildfires has always been a herculean endeavor as well as an apparent mismatch—perhaps the ultimate example of man battling with nature and often hoping for nothing more than a tie. Containment is the primary goal, as opposed to extinguishment—minimizing encroachment into heavily populated areas by dousing the fire with liquid or chemicals and utilizing squadrons of ground crews to build fire breaks that encourage nature to run its course as benignly as possible until eventually the fire runs out of fuel and dies.

It is a process that occasionally, in the case of small brush fires squelched quickly with water and choked by friendly weather conditions, takes a matter of hours or days. More often, and increasingly in this era of ceaseless unpredictable wildfire seasons, smallish fires erupt into conflagrations that surge and spread for weeks or even months, scorching tens of thousands of acres of wildlands, disturbing fragile ecosystems, and, with alarming regularity, threatening densely populated urban and suburban neighborhoods. The Camp Fire burned more than 150,000 acres and killed eighty-five people in November 2018, virtually wiping the town of Paradise, California, off the map in the process. At almost the same time, a few hundred miles to the south, the Woolsey Fire ripped across Ventura and Los Angeles Counties, devouring nearly

100,000 acres before finally running out of fuel at the edge of the Pacific Ocean in Malibu.

Wildfires, as devastating and terrible as they can be, are part of the natural order of things, like hurricanes in the Gulf and tornadoes in the Midwest. They are, in some sense, as normal as the sun rising and setting, a part of the cycle of life and death in the wilderness. But they are no longer restricted to wildlands, and the death portion of the cycle now threatens to outweigh the portion devoted to life. They are burning hotter, faster, longer, and more ferociously than previously seen, coupled with an increased human interface and fueled by a changing climate that is more welcoming to the severe conditions in which fire thrives. So, management is crucial, facilitated by a broadening of resources and an ever-expanding array of technological innovation. At the heart of the fight, though, are the men and women who march into the flames or pilot the aircraft that provide support. And the Super Scooper, an aeronautic marvel whose mission and skillset are seemingly a throwback to a simpler time, is perhaps the most valuable tool in the kit.

Bridger Aerospace is the first and only U.S. company to own and operate the CL-415EAF, and we are eager to baptize them under fire, to demonstrate that they are not only a sound investment on the part of our financial backers, but an extraordinary aircraft that should be on the front lines of the battle against wildfires. It's a bit of a joke in the industry—the way the media gets orgasmic over the sight of a tanker dumping retardant over a majestic, fire-tipped tree line, plumes of neon pink streaming from its underside. It *looks* fantastic! No question about it. And indeed, tankers have been the backbone of the American aerial firefighting effort since their inception in the mid-twentieth century. But the rest of the world uses a scooper–centric strategy that emphasizes initial and direct attacks to suppress the fire immediately. The U.S. has embraced a retardant–centric strategy of indirect attack, management, and containment. A valid strategy, but an incomplete one considering the vast size and diversity of the American wildfire geography. With nearly two hundred scoopers operating worldwide and a grand total of four working in the U.S. as we prepared to enter the market in 2020, it was time for America to get her foot into the direct attack game!

There is obviously a critical role filled by the old-fashioned jumbo tanker—usually a retrofitted commercial jet—in the effort to contain wildfires. But the truth is that tankers loaded with retardant carry an unseemly set of side effects that many in the industry and government prefer not to talk about: ground and surface water contamination by retardant-based chemicals, corrosion of airframes by the chemicals, health side effects to ground firefighters breathing the vaporized gas of burned retardant, and, of course, the impact on wildlife of an unnatural chemical mix dropped in their midst. There is no question that the use of retardant is almost always warranted and the "right answer." But with America's willful ignorance of scooping aircraft, the question must be asked: How many times has retardant and containment been used when a direct attack water-based strategy could have done better?

The Super Tanker, a converted 747, can carry up to twenty thousand gallons of water or retardant. More commonly used are slightly smaller Type 2 or 3 tankers that carry around two to three thousand gallons of liquid. While both aircraft have significantly greater payload capacity than the Super Scooper, they are limited by two factors: they drop their loads from a greater altitude, which makes them less precise, and they must return to their home base to reload after each drop. Depending on the location of the fire and its proximity to a fire base, a Super Tanker might have to travel hundreds of miles between drops. Time spent traveling, refueling, and reloading naturally reduces the efficiency of the tankers.

The Super Scooper, on the other hand, carries a relatively modest fourteen hundred gallons of liquid, but its ability to reload on the fly (so to speak) and deliver precisely targeted low-altitude drops—dozens in a single day without refueling—makes it an impressively efficient and economical aerial firefighting machine.

Or so we've heard…and have come to believe based on training exercises. But there is nothing like a live event, an honest-to-goodness wildfire, to put theories and practice to the test. And while it may sound strange to wish for such a thing, we long for the opportunity. This is in no way a desire to court destruction; wildfires exist, and our job is to fight them. They are coming—inexorably, inevitably—whether we like it or not. All we can control is our response. Typically, scoopers are less utilized in "project fires"—those that have been burning for weeks or months. They are more commonly employed in initial attacks, when a fire is still new and perhaps manageable—when there is still the possibility of smothering the blaze with water before it gets out of hand. As such, scooper missions are sudden, unpredictable, and subject to last-second cancellation. We had experienced a couple false starts in the previous two weeks, times when a call came in and we were told to be ready, so we fueled up the planes…and then we sat. And sat. And sat some more.

Sorry, false alarm. We don't need you today.

The truth is that you are never truly on a fire until you are *on a fire*. But once you engage, there is nothing quite like it.

★ ★ ★

Finally, on the morning of July 20, the call comes. There is no warning, no buildup, no sense of anticipation fueled by a nearby major event that has raged for weeks. Instead, there is a spark of unknown origin and a fresh new fire located near Elko, Nevada, some two hundred fifty miles northeast of Minden, not far from the Utah border. But we don't even know this much at the time. Instead, we are hanging out at the airport, so bored that we have chalked up a section of blacktop outside our trailer to mimic a foursquare court, and now we are batting a ball around—grown men engaging in a schoolyard game of their youth just to pass the time. Like crews in an urban firehouse, we are simply waiting for the alarm to sound.

Suddenly, a cell phone rings. I look over and see our chief pilot, Tim Langton, listening intently.

"Dispatch," he says silently, mouthing the word.

Within a minute we are scrambling. I'm the CEO, the "boss," but in this scenario, I take a back seat, literally and metaphorically. The Super Scoopers are new to the U.S. market but have been in use in Canada for a while, so as part of our program, we've brought in experienced Canadian pilots, like Tim, to serve as captains and help train some of our pilots who have experience flying air attack but have little firsthand experience with scoopers. ("Air attack" is a term used to describe a plane that is flying high above a fire, basically controlling traffic—the delicate dance performed by a fleet of tankers and scoopers and helicopters that dump water, flame retardant, and even smokejumpers onto a fire.) I have trained on a scooper, but I've never flown "live," so on this mission I climb into the jump seat, between and behind Tim and another captain, Paul Evans, a career Scooper pilot from Ontario. I'll man the radio, which is a complicated bit of choreography all its own, while they focus on flying the plane.

On the roughly ninety-minute flight to Elko, I pull up maps and GPS coordinates of the area, and I begin looking for bodies of water in proximity to the fire. As it happens, we're in luck—the South Fork Reservoir, easily accessible and roughly three miles long by one and a half miles wide, is only five miles from the burn. As yet, that is all it is, just an unnamed fire event of undetermined origin or ferocity; by the end of the day, fed by strong winds and arid conditions, it will have been dubbed the Cedar Fire and consumed some three thousand acres of grass and timberland. When we arrive, the only other aircraft on scene is a lone air attack Twin Commander (the staple air attack platform of the aerial firefighting community; Bridger has owned ten of them), circling overhead, preparing to manage the Fire Traffic Area (FTA). We are the only suppression aircraft. I check in with him fifteen miles west-southwest of the fire.

"Cedar Creek Air Attack, Scooper 281 checking in. Four hours fuel, three souls aboard, we have the fire in sight. Ready to go to work."

"Roger...uhhh. Who is this again?"

"Scooper 281, dispatched out of Minden."

"Copy. Uh, what kind of scooper? Who you guys working for?" He wanted to know what agency had sent us....

Not exactly the welcome I expected, but we would get used to it; as the new kids on the block in aerial firefighting, we would be asked every time we showed up—usually with a tone of skepticism or annoyance—"Who the hell are you?"

"Uh, well we are a CL-415EAF Super Scooper. Under contract with State of Nevada."

"Huh. Okay, well that's cool. You'll see me at your two o'clock high in a right-hand orbit. I want you working the south flank of the fire and keep it from crossing the drainage. Scoop site will be South Fork Reservoir, left hand pattern off the scoop, left hand turn off the drop. Let's get to work."

And like that, we are in business. We head to the reservoir, just a few minutes away, and begin our first approach.

It's hard to convey the drama of this initial drop: the thrill of christening such a magnificent piece of machinery in service of a noble effort, the tension that comes with executing a maneuver that has zero margin for error, the insistence upon calm professionalism in a scenario that is, by definition, chaotic. The Super Scooper's normal cruising speed is two hundred miles per hour; while loading, however, the aircraft is slowed to a speed of approximately one hundred miles per hour. As we approach the reservoir, dipping lower and lower, a pair of probes descend from the bottom of the plane, one on each side of the keel. Together, they form a shovel, less than ten inches wide, that funnels water into a pair of tanks in the belly of the plane. It is an extraordinarily delicate exercise, the pilots gently feathering the controls to skim the water at precisely the right depth—too high, you pick up nothing; too low, well...you crash. And weather conditions—turbulence, wind-driven whitecaps on the water, low visibility—all complicate the procedure.

Not surprisingly, there is also virtually no conversation in the cockpit outside of that which pertains directly to the mission at hand. This is partly due to the seriousness of the job, but also in keeping with FAA guidelines, which mandate a "sterile cockpit" during the critical phase of any flight—commercial, charter, or cargo. Critical phase is generally defined as within ten miles of an airport or at an altitude of less than ten thousand feet, the parameters established because this is where 95 percent of all aviation accidents occur. It is also where Super Scoopers spend most of their existence (other than when flying to or from a firebase).

In precisely twelve seconds, the scoop is over, the drag on the plane relents (there is a pronounced feeling of release as the probes retract), and we begin to ascend, the aircraft now heavier by a third. The simple and irrefutable rules of physics and aerodynamics would seem to indicate the folly of such an ascent, particularly at this low rate of speed, but again, the Super Scooper is engineered for exactly this type of exercise. With the heft of a C-130, it may not look like a stunt plane, but, thanks to its supersized flight-control surfaces (ailerons and rudder), it is in fact a nimble plane capable of surprisingly aerobatic maneuvers, such as darting though narrow spaces and banking high-G turns that would pull a lesser-equipped aircraft to the ground. Once we've filled its tanks with water, the Super Scooper's massive turboprop engines, generating 2,380 shaft horsepower per side, hoist the plane into the air. The extreme curvature of the wings (known as camber) provides additional lift.

And just like that, we are on our way, an aerodynamic marvel—a giant flying tank with a belly full of water, cruising to the fire.

"Scooper 281 inbound," Tim reports. "Off the scoop."

Within two minutes, we are over the fire, which is burning fiercely on the side of a mountain, both wind and terrain driving it southbound. We come in from the high side, following the fire down the slope, and release our payload, while cruising at a speed of 110 knots (125 mph). There are no ground crews on-site yet, so we don't have to worry about hitting anyone (fourteen hundred gallons of water, if accidentally dropped on human firefighters, could have fatal consequences). Still, protocol dictates a verbal accounting.

"Scooper 281, line is clear, clear to drop."

"Roger, 281 clear to drop."

The water splashes below like a water balloon on a bonfire. It's hard to know in that moment what effect the drop has had, but we know that it was successful. We also know that it will be the first of many on a very long day.

As soon as the drop is complete:

"Scooper 281, off the drop."

"Roger 281, good drop, load and return, make that your pattern for the rest of your fuel cycle. We will have two Type 1s [helicopters] inbound; we'll reassess your pattern when they arrive. Good to have you."

After all the years of buildup, the fundraising, the drama of getting certified and carded, the plane crashes, the near-death experiences… finally dispatching to our first fire is an amazing experience and an incredible rush. But, not unlike my first gunfight overseas after many years of intense training, the bullets don't know any different. It is oddly anti-climactic. The air attack coordinator overhead who is calling our drop doesn't seem to know or care that it was our first mission in the U.S. and that we are on a very odd state contract that puts severe limitations on our ability to operate. We are simply another plane in the fight; we dropped where he needed us to drop, and we did it well. He put us in the pattern and treated us like any other asset. And as the day wears on and more assets arrive on the scene, we become the lead aircraft since we were on the scene first and know the terrain, the scoop site, and the scenario.

Lap after lap, we fly to the reservoir, scoop, then return to the fire and await clearance to drop. With each trip back, the sky is more crowded—helicopters, air tankers, air attack, and finally three more scoopers from another company. The fire spreads quickly in the first few hours, the sagebrush and grass and trees igniting like tinder in the hot July sun of northern Nevada, but by late afternoon, containment is within reach as we basically pound the flames into submission. We complete seventy-nine drops that day, more than 120,000 gallons of water discharged from a single aircraft.

Over the next three days, the fire lingers and fans out to some six thousand acres, but a vast constellation of resources—some 260 personnel including ground crews—contains the fire with admirable efficiency. To put this in perspective, the entire island of Manhattan is fourteen thousand acres. So a fire of this magnitude, left unchecked, could basically destroy an area the size of New York City in a matter of just a few days. It is a vivid example of the effectiveness of aerial firefighting and how employing a diversity of tools—without hesitation—creates the most positive outcome. For all the frustrations of interagency bureaucracy and climate change and urban-wildland interface, not to mention wildfires getting worse and more persistent by the year, there is cause for optimism. After years of bureaucratic obstacles slowing and frustrating our progress, we finally see some light at the end of the tunnel. And just in the voice of the air attack that day, we are once again reminded that although "the bureaucracy" can be intractable and terrible, it is generally filled with good people who want to get the job done. And if given the opportunity, they will always try to do the right thing for the people on the ground.

There is, at the very least, hope.

CHAPTER 2

AMERICAN MUDSLINGERS

The following is a story of courage, grit, innovation, desperation, lies, deceit, love, and competition.

From the early days of the B17 to the modern fleets of the twenty-first century, we will take you on a ride through the history of aerial firefighting—the most hazardous and demanding aviation mission in the world—as seen through the eyes of someone who has both studied its evolution and fought on the front lines. Part memoir, part narrative history, *Mudslingers*, I hope, will be a rollicking read, an enlightening journey, and a call to action for anyone who believes wildfires are not only one of the greatest threats facing modern civilization, but a threat that has long been underestimated, misunderstood, and poorly addressed, despite repeated examples of bravery and innovation by those who choose to do battle with the flames.

This book, in gestation almost from the first time I saw a wildfire up close, is a humble attempt to chronicle the history of aerial firefighting in a brief and digestible format. Inevitably, it is also a story of how I became involved in a world that I knew almost nothing about before joining the fray. As a lifelong pilot, military veteran, history enthusiast, and student of all things related to aviation and aerospace, it is reasonable to assume that if I knew next to nothing about the history of firefighting aviation, very few others did. And therein lies my mission: to broaden the awareness of aerial firefighters by bringing their story to light while also sharing my own experiences. I have left so much out of this story that simply couldn't fit within the pages of this book. There is a reference section at the end that provides some additional technical details and narratives to fill in some of the items that could be pieced into the story in this book. There are heroes of aerial firefighting that are not in these pages, and for that I apologize. It is not a comprehensive history, but rather an attempt to share the basics of the history of the American Mudslinger with as many Americans as possible.

It is also worth noting that 100 percent of author proceeds from this book will be donated to the Montana Firefighter Fund, the Associated Aerial Firefighters Foundation, Wildland Firefighter Foundation, and the United Aerial Firefighters Association (UAFA). These four organizations have a direct mission of helping the families of wildland and aerial firefighters who have been injured or lost in the line of duty. They are the reason that we are telling this story.

Never has this topic been more relevant and critical than it is at the time of this writing. The entire world faces a growing wildfire crisis—from Australia to Sweden, Chile to Alaska, Brazil to Thailand, and almost everywhere in between. As these fires escalate in speed, severity, size, and frequency, aerial firefighters and their aircraft have become ever more crucial to the global populace. Our governments—national, state, provincial, local—and even international cooperation organizations, like NATO and the UN, must engage with firefighting more seriously to better manage and suppress the international wildfire crisis we are facing.

The spring of 2023 brought unprecedented wildfire activity to Canada, with fallout felt across vast portions of the Eastern Seaboard of the United States, a region not ordinarily accustomed to dealing with severe wildfires. By early June, however, millions of acres of forest in nine of Canada's thirteen provinces had been devoured by extraordinarily aggressive fires. While Canada is certainly no stranger to wildfires, the heaviest activity historically has been in the western provinces; cooler, wetter conditions in the east typically provide a less welcoming environment—particularly in the springtime. As of this writing, however, much of the country is under siege, with aerial firefighting crews from around the world dispatched to even the eastern provinces, like Quebec and Nova Scotia, where plumes of toxic smoke from wildfires drifted many hundreds of miles to the south, blanketing U.S. metropolitan centers from New York to Washington, D.C., and leaving skylines bathed in an eerie orange glow.

If ever there was a moment to recognize that wildfires had become a ubiquitous threat, surely this was it. And while climatologists, environmentalists, and other scientists can argue over the root cause, and the extent to which climate change plays a role in increased wildfire activity, this much is undeniable. Wildfires are getting bigger, faster, stronger, more deadly. And we need to get serious about how we are going to fight them in both the long and short term.

But, as with most global challenges, the truth is that we cannot effectively pressure our elected officials to act unless we have at least a rudimentary understanding of the problem and how to fight it. In the twenty-first century especially, we are overwhelmed by so much information (a significant portion of which can be categorized as *misinformation*) that it can be daunting to discern what is truth and what is fiction. Having a historical perspective on issues has always helped me to craft solutions to better address those same issues as we look into the future.

I am an accidental aerial firefighter. It was never something I had a desire to pursue simply because, for most of my life, I had no idea the profession even existed. When I started Bridger Aerospace after I was discharged from the military in September 2014, fighting wildfires was not a consideration. But the circuitous route that led me to firefighting perhaps helped make me an ideal messenger for this book. Many stakeholders in this industry (and, yes, fighting wildfires is indeed an industry, one that is, by necessity, growing by leaps and bounds) have been in it or around it for decades—in many cases, generations. Fathers and grandfathers flew on fire or fought it on the ground, and the communities that engage in these battles or have been destroyed by fire bear scars and the strong opinions that come with them. Predictably, these scars warp the perception and objectivity of the wildfire community and its history in such a way that it can be hard to get an honest account.

My first experience with fire aviation—indeed, with wildfires on any level—was in July 2015 (I was twenty-nine years old); all my previous professional flying experience had been in the United States military. I loved it from the very beginning. It was true love, as opposed to infatuation, because the more I saw and learned and experienced, the more deeply I fell in love. And as my business enterprise grew and I became more personally enmeshed in the world of aerial firefighting, the more I found myself fascinated by its history: the people, the fires, the planes, and the agencies behind it all.

Several years removed from my last college class at the U.S. Naval Academy, I became a student all over again. I read all I could get my hands on about wildfires, the people who fought them, and the technology they employed. But try as I might, I could not find a seminal work that captured the history of aerial firefighting in a condensed manner. Those that existed were either reference books with a technical focus, novels with a historical twist, or narrow looks at the industry, usually embedded within a larger story about the phenomenon of wildfire. That said, I am indebted to the many authors who came before me in chronicling much of the amazing history of fire aviation and of American wildfires in general, most notably in *The Big Burn*, *Young Men and Fire*, and *Aerial Firefighting*.

Wildfires are a global threat, and there are brave men and women around the world who conduct our mission with skill and tenacity. However, this book is largely an American story with some international vignettes. Perhaps a later edition can address the international aerial firefighting market, but since aerial firefighting started in the U.S. and Canada, and it's where most of the innovation thus far has occurred—and it's where my personal experience lies—I will focus primarily on that story. I mean no disrespect to the brave aerial firefighters of Europe, Australia, Asia, or South America by omitting most of their history.

I will also freely state up front that many fascinating and critical parts of aerial firefighting history have been left out of this book. It was a challenging and at times emotionally wrenching experience deciding what vignettes to tell, what people to highlight, what fires, what planes, bases, companies, and so forth. Our task was to put together a story that was not just accurate, but also digestible and, hopefully, entertaining. This necessitated many hard decisions that will undoubtedly leave some paramount figures feeling slighted; I apologize in advance for this and only hope you can forgive me long enough for a more comprehensive history to be published. My goal in this volume was to pair a brief but compelling history of this remarkable industry with my own story and my own view—sometimes from ground level, sometimes from within the clouds.

At its core, this is a story of war: The wars in Iraq and Afghanistan where, as a Navy SEAL, I learned how to lead, how to follow, to fight, and to learn. The war against bureaucracy in American government, the military, and wherever it dehumanizes a situation and creates unnecessary suffering as a result. War against complacency. War against the corrosive nature of the status quo. And most importantly, war against wildfire. Because that's what we are in—a brutal and unending war. And unfortunately, in any war, as an intractable bureaucracy contemplates and considers and considers, young men and women die on the front lines. Millions of global citizens are displaced, homes incinerated, communities destroyed, ecological domains forever altered. In the summer of 2023, State Farm and Allstate stopped writing new homeowners insurance policies in California. That is a huge impact on homeowners; after all, without insurance, you can't get a mortgage, home equity loan, or rent your home.

In 2023, the year of this writing, you cannot talk about wildfires without talking about climate change, or lack thereof. This is a highly politicized and divisive subject, and the degree to which one believes in the fearsomeness of climate change often comes down to how far left or right they stand on the political spectrum. That is a shame, because climate change is neither a fantasy nor merely a political tool, but it is getting increasingly challenging to see the issue through clear eyes. Whether you believe it is caused by cars, planes, gas, diesel, cow farts, solar radiation, orbital dynamics, propane, methane, butane, or the sabertoothed tiger, the climate is always changing and will always be changing, and our small part in it is to fight the fires when they come.

But, you might ask, isn't wildfire *natural*? Isn't it supposed to occur in our ecosystems? The clear and obvious answer is an emphatic *Yes*! Fire is a critical component of almost all natural ecosystems. It helps to rejuvenate forests, replenish grasslands, restore chemical and nutritional balance to soils and plant life, and, when properly sequenced, reduces the risks of future catastrophic wildfires. But as with so many things related to nature, wildfires are more complicated than they might appear at first glance.

Different ecosystems are meant to burn at different times of the year, at different temperatures, and at varying frequencies. For example, a heavy timber forest of Douglas Fir in Northern Idaho is, by nature's standards, supposed to burn once every five to eleven years, usually sometime in July, August, or September. Theoretically, because the last fire burned this patch of forest just a few years prior, and the moisture content of the soil, trees, and pine needles is relatively high in this region, the natural fire will likely burn through the forest floor. This means it will clear invasive brush species, kill off newer saplings that may be invading the space of other trees, replace critical nutrients in the forest soil, and most importantly, it will clear out dead and dying branches, felled trees, and underbrush that can suffocate a healthy forest. This deadfall prevents free movement of forest creatures that help keep nature's balance, prevents enough moisture and sunlight from reaching the floor of the forest, and clears out the lower portion of tree branches that allow this "under-canopy" area to remain open and healthy.

In fact, for thousands of years, Indigenous tribes on all continents would use fire to efficiently manage their native wilderness. This was an effective and cheap way to keep their most precious resource—land—healthy and productive. Many things have changed over time, particularly in the last hundred years. Wildland–urban interface, drought, climate change, "active" forest management that is anything but active—all these things, and many others, have contributed to a wildfire season that is no longer a season but a year-round battle. The tools and technology we have at our disposal in fighting this war have improved exponentially, but the opponent has never been more fierce, more destructive, or more terrifying.

It's important to note, however, that there is a growing divergence in the scientific community about the veracity of human-caused climate change. In some circles, it has become a sacrilege to even question the conclusions of the UN's Intergovernmental Panel on Climate Change (IPCC). But it is core to the very nature of science to question everything. The simple truth is that the Earth's climate has been in a state of constant change since the beginning of time. From the Cretaceous period, when the average global temperature was 10 degrees Celsius warmer than it is today, to the "little ice age" of the eighteenth century, we have seen various periods of natural climate change that had nothing to do with cattle flatulence or diesel emissions. Either way, when a wall of flame is threatening your home, your family, your livestock, or your life, whatever global weather pattern is causing the fire is not your concern, nor is it mine as I line up for a water drop. The mission is to stop the fire and protect the people on the ground.

★ ★ ★

Aerial firefighting is a relatively new tactic in man's attempt to harness nature's fury. For thousands (millions!) of years, nature took care of itself, and humans could do little but watch and hope. The occasional controlled burn by Indigenous tribes was not merely the most logical approach to mitigating the impact of wildfire, but the *only* approach. With the increased colonization of North America and the subsequent expansion of populations into areas once deemed practically uninhabitable, wildfires became not just more common, but bigger, stronger, and more formidable.

From the middle portion of the nineteenth century, with the great western expansion, American civilization began encroaching upon areas that were not merely subject to the threat of wildfire, but in fact were *meant* to burn with some regularity. Thus was born the concept of fire suppression, a somewhat quaint and naïve notion that man has the power to tame nature's wrath. In the days before the advent of flight, this amounted to utilizing teams of men armed with shovels and rakes and hoses attached to mobile ground tankers—in some ways, it was no different than fighting fires in an urban setting. Douse the flames with as much water as possible while simultaneously trying to steer the course of the fire in a direction that will do the least amount of damage to humans, livestock, and property.

When the entire fight is being waged from the ground, it is a decidedly one-sided affair. But the battles were waged nonetheless, with varying degrees of success. Fire suppression became a double-edged sword, even in those early days, as fires seemed to increase in severity. This is the great paradox of fire suppression and a quandary faced by firefighters even today: putting out the fire—every fire—as quickly as possible might seem like a noble objective, but there are consequences to suppression that must be faced down the line, most notably in the form of accumulated material that is eagerly waiting for a spark. Indeed, the term "fuel load" is one of the most debated in the world of firefighting because it refers not just to burnable land masses—whether urban or rural—but to an aggregation of combustible material that in previous generations (or centuries) would have been extinguished through natural means over time.

Simply put, the quicker you extinguish a fire, the more likely you are to have to deal with an even bigger fire at some point in the future. This is not to suggest that every fire should be left to chart its own course and burn to its logical conclusion. Obviously, that would have catastrophic consequences, particularly in populated areas. But the fact remains that massively increased fuel loads have changed the calculus of wildfire management. Controlled burns have their place in this discussion, but they are not without risk for the same reason that natural and uncontrolled burns are dangerous: wildfire is powerful, unforgiving, and unpredictable.

It is, after all, *wild*.

"There is no location in the U.S. that's remote enough to say it's a good idea to let the fire burn," noted John Gould, the president and CEO of 10 Tanker, an aerial firefighting organization whose primary asset is DC-10 air tankers. Gould was also a ground firefighter, smokejumper, and air attack supervisor for many years, so his breadth of knowledge and experience is impressive. "There are fires that burn from Bend, Oregon, all the way over to Salem. There are fires that burn over the Sierra Nevadas. In Alaska, Fairbanks is always threatened; Anchorage is threatened; every village in Alaska is threatened. It's just never a good idea to let the fire burn. What really has to happen in this world that we live in today is you have to say, 'All right, from May 1st to say, October 1st, there are no good fires.' Regardless of what we tried to make people believe—that we can fireproof the forest, or that we can change the dynamic somehow—if you really want to save your forests, you put every fire out as fast as you can. There's no fire that's a good fire."

Consider the Calf Canyon/Hermit's Peak fire that destroyed more than three hundred homes and displaced thousands of residents in Northern New Mexico in the spring of 2022. Both Calf Canyon and Hermit's Peak fires were prescribed burns supervised by the U.S. Forest Service (USFS) with obviously good intentions. Both fires, however, jumped their containment lines and eventually merged to form a mega-fire—a "project" or "complex" fire, in industry parlance, so-called because, indeed, it is a project and often requires weeks or even months to complete with the aid of firefighters on contract with multiple state and federal agencies. It remains true that prescribed burns rarely get out of hand, but it is also true that project fires have become much more common, and the damage they can inflict is both awesome and devastating. The Calf Canyon/Hermit's Peak fire became the biggest fire in New Mexico history, burning more than 300,000 acres. A fire that first sparked in late January was still less than 50 percent contained on Memorial Day weekend.

This does not mean that all prescribed burns are a bad idea or should be abandoned entirely as a means for coping with wildfires. But it is an indication of the bureaucratic, logistical, and ultimately human risks involved. Increased fuel loads make it imperative that prescribed burns be utilized in some capacity; at the same time, an ever-smaller window for completing prescribed burns, combined with climatological forces that can whip the smallest of fires into a project that threatens life and property, makes it harder to sell the concept of prescribed burns to a wary populace—and to the government officials who will take most of the heat (so to speak) if things go awry.

The tension between local, state, and federal agencies and private landowners has grown significantly in recent years as the wildfire crisis has worsened. Staffing challenges and the perception of federal response as being too slow is making cooperation harder as strategies conflict. For example, in October of 2022, during a prescribed burn in Grant Country, Oregon, the fire spilled over the planned fire line and burned twenty acres of private land. The prescribed fire itself was the subject of much controversy prior to the burn. As tensions came to a boil, the local sheriff was called to arrest the duly appointed federal incident commander, Rick Snodgrass, with charges of arson. Rick was arrested and charged, precipitating a tense standoff between the U.S. Forest Service and the Grant County Attorney's office, setting the stage for what will be a seminal case in the future of American wildland fire. As of this writing, no trial or settlement has been announced. This incident was the first of its kind and sent shockwaves through the wildland fire community and is sure to usher in a new era of federal and local tension around fire incidents.

To adequately confront the future of fighting wildfires, it helps to understand the past. No history of wildfire is complete without first discussing the Big Burn, also known as The Great Fire of 1910, the largest wildfire in American history, and most likely one of the largest in the history of the world. A combination of electrical storms that ignited brush fires, unusually dry late-summer weather, and hurricane force winds combined to produce a natural disaster of biblical proportions. The Big Burn, a collection of some three thousand fires, torched more than three million acres in Montana, Idaho, and Eastern Washington. The bulk of the fire lasted just two days before being doused by heavy rains; most of the damage occurred within a tornadic period of six hours during which, according to observers, entire mountainsides erupted into flames, and sixty-foot trees were pulled from their roots and fired across canyons like giant flaming rockets.

The Pulitzer Prize-winning journalist Timothy Egan exquisitely chronicled the story of the Great Fire of 1910 in his bestseller *The Big Burn*, which remains the definitive history of the event. Egan deftly wove the story of the fire and those who fought it with the efforts of Teddy Roosevelt, an outdoorsman who became president, to build support for the notion that public land was a national treasure to be protected and enjoyed by citizens throughout the country as opposed to a privileged few.

During the decades spanning the end of the Civil War to World War I, much of the American West was beset by corporate greed, which was as much a threat to the safety and security of settlers as clashes with natives had ever been. The entire region north of Salt Lake City, west of Rapid City, and east of Spokane was little more than a privately owned preserve of logging, mining, ranching, and railroad interests. Little happened there that wasn't the express desire of East Coast titans, wielding more money and influence per capita than any other corporate barons in the history of mankind—even more than the tech titans of today! For although their net worth may have been less, their grip on the levers of power was tighter than any CEO today could ever hope to achieve. It was such a toxic situation that Roosevelt spent half his presidency fighting to end it.

Montana, barely even a state at that point, was one of the reasons U.S. Senators became directly elected by the people. Prior to the Seventeenth Amendment in 1913, senators were elected by the legislature of each state; to become a senator, all you had to do was convince fifty-one out of one hundred of your aristocratic peers that you were fit for office, and that was that. This started as a good check to ensure that senators served as ambassadors of each state to the president, and the appointees were insulated from the need to campaign and partake in partisan arguments that were considered beneath their office. But eventually it became a corrupt system in which it wasn't hard to imagine the owner of a prosperous company in, say, Butte, trying to convince the legislature that it was in their best interests to appoint him senator, lest he decide to flex the muscle of the Anaconda Copper Mining Company and slowly but firmly choke off the economy of the nation's most wild state.

For the purposes of our story, it is worth revisiting the Great Fire of 1910 and its political and socioeconomic aftermath simply because it fundamentally impacted U.S. Forest firefighting policy and influenced forest management to the present day. It is not an exaggeration to say that the Great Fire of 1910 laid the foundation for the creation of modern-day American wildland fire management, which largely served as the model by which other nations eventually managed their public lands as well. And, although it could be argued that the U.S. eventually took a back seat to other nations when it came to aerial firefighting, it no doubt was at the forefront of the revolution. Of course, no discussion of the 1910 fire is complete without the appropriate acknowledgement of Edward Pulaski (the namesake of the famous Pulaski tool, the half ax-half pickax tool that is the primary weapon of every wildland firefighter) who, through quick thinking and remarkable courage, saved dozens of lives by hiding out in a mineshaft during the inferno. His name lives on today as the cultural touchstone of American wildland firefighting.

★ ★ ★

Fighting wildfire is no different than any other tactical operation. Present a certain scenario to ten different tactical incident teams and you will get ten different approaches to handling the situation. In general, there will be fundamental tenets that no one should violate, but all professionals are victims of their own training and experiences. A firefighter who is deeply experienced in combating the intense timber fires of the Northern Rockies, Pacific Northwest, or British Columbia will have different opinions on the use of heavy equipment, aircraft, and ground crews than an incident commander whose base of expertise is in fighting the fast-moving grassfires of Texas, Oklahoma, or Central California.

Regardless of background, firefighters generally work under the auspices of American federal authorities who are guided by a wildland fire *management* philosophy, meaning they view wildfires as incidents to be "managed" in much the same way that one would manage a plan to control an invasive species or a crop rotation program. There is sound reasoning behind this philosophy. Fire is a critical part of our natural ecosystem, and at some level, almost all nature in America—be it grasslands, forests, or sensitive riparian habitat—is meant to burn every few years. Given the fact of this natural cycle, when a fire breaks out in, say, Glacier National Park, to immediately treat that fire as something that must be immediately eliminated is neither ecologically wise nor especially practical. In truth, that natural habitat must burn and, in fact, is probably several years overdue to be burned. Furthermore, considering the remote nature of many of our natural habitats, responding rapidly with fire suppression resources may not even be possible—it could be hours or even days before the proper assets are available to address such an incident.

The incident commander assigned to this fire, most likely a local forest ranger whose primary job is overseeing visitor centers and trail upkeep, must now balance the desire to limit the potential damage of the new start against the ecological needs of the park area. Additionally, the incident commander must be budget-minded, since he is not authorized to simply press the proverbial red emergency button and summon every potential firefighting asset imaginable. Understandably, any citizen who resides near a wildfire assumes that whenever a new fire is detected, the "red button" is automatically pushed and resources from every corner of the nation sprint toward Glacier National Park or Yosemite or a suburb in Southern California. In truth, the prioritization of these incident—and the type and extent of resources deployed—is far more nuanced, complex, and, frankly, political than one might imagine.

When a new start is reported, an evaluation (colloquially referred to as a "size-up") occurs almost immediately. As the name implies, a size-up entails assessing the initial scope (or size) of the fire, the direction it's traveling, fuel type and disposition, wind direction and weather conditions, and of course, the likelihood of danger to structures and people. This basic size-up produces an initial request for assets and a determination of the urgency related to those assets. The incident commander may determine that the new start has begun in an area of dense timber with significant fuel available and with wind and terrain favoring the movement of the fire, and thus likely to become a very active and intense blaze quite quickly; however, if there are no people or structures within, say, fifteen to twenty miles of the new start, and it is limited to a wilderness area, a decision might be made to simply "observe" the fire before any assets are called.

This initial evaluation can occur by various means. Sometimes, it's the ranger himself, driving out into the mountains in a pick-up truck to gain a vantage point. Other times, it's a pre-positioned camera on what used to be a manned fire tower, recording and sharing real-time video footage. Most common, however, is the use of an aerial reconnaissance aircraft (also known as an air attack aircraft). These birds are assigned to fly a region of the country and sometimes act as spotter aircraft for "new starts" (the term refers to a newly ignited blaze that can pop up in any given area) although their primary purpose is to serve as the aerial command post during a fire.

Once the aircraft assigned to the area provides the incident commander with a size-up, he and the forestry professional in the aircraft together request resources from the regional dispatch center or Geographical Area Coordination Center (GACC). This request could include ground resources, such as bulldozers and hand crews, and aerial resources, such as single-engine air tankers, dedicated air attacks, and large air tankers. This request is not received in a vacuum, obviously; there are not limitless resources to fight wildfires, and the resources that are available typically come with a steep price tag. If the incident at hand is a new start in the middle of Nevada on land owned by the Bureau of Land Management (BLM), the largest owner of land in the United States), and that fire does not have a single inhabited structure for one hundred miles in any direction, the GACC may very well decide to send no assets to that new start. It is government land that is meant to burn periodically. Right or wrong, budgetary considerations play a major role in determining the response to a new start, and the GACC rarely sees the benefit in blowing its budget on fighting a fire in the middle of nowhere.

Complicating this process is the BLM's long-standing practice of allotting some 20,000 permits to ranchers whose livestock grazes on 155 million acres of public land—land that is sometimes "meant to burn." To those ranchers, understandably, a fire "in the middle of nowhere" can be devastating.

Far less arbitrary and ambivalent are the decisions made when a fire breaks out in a place like the foothills of Los Angeles County (as was the case with the Woolsey Fire), which has a population of ten million people. When lives and property are so obviously and imminently threatened, urgency rules—and suppression, rather than merely containment, is typically the goal. The extent to which that goal is attainable depends on a multitude of factors, not all of them controllable or even foreseeable. But as with most wildfires, it starts with a quick response and the deployment of an effective aerial firefighting force.

From the very beginning of an incident's lifecycle, aircraft are a critical piece of wildfire management. When a fire spreads over tens of thousands of acres, observing or controlling that incident from a ground vantage point, no matter how high or well placed, becomes nearly impossible. And in the case of fires in more heavily populated areas in which suppression is the objective, tankers and scoopers become the first line of defense. Simply put, an aerial component at least gives humans a fighting chance in a battle against one of nature's most overwhelming forces.

As the fires of the American West moved out of the mountains and began threatening communities in the rapidly growing towns and cities of the Pacific Northwest, California, and British Columbia, the use of aircraft to help manage the blazes became critical. There were pockets of innovation all over North America that helped to effectively integrate aircraft into the wildfire management paradigm. The earliest pioneers centered around a handful of locations to include Sault Ste. Marie, Ontario; Chico, Hemet, and Grass Valley in the San Joaquin Valley of Central California; Missoula, Montana; and the area around Vancouver, B.C. There were others to be sure, but no one can deny these locations were some of the most fertile breeding grounds for firefighting aviation.

Understandably, the early role of aircraft in fighting wildfires was restricted primarily to the observation of forested areas—basically providing a much-needed bird's-eye view of fires as they sparked and spread across a wooded region. This allowed for improved strategizing and smarter and safer utilization of ground forces in managing wildfire. The earliest recorded usage of aircraft for this purpose was in 1919, when the U.S. Forest Service incorporated several small aircraft, including the Jenny biplane, to patrol vast stretches of wilderness in the Pacific Northwest. While limited in utility, the Jenny and its siblings nonetheless represented a great leap forward in the effort to manage the threat of wildfires. Manned fire towers had long been utilized to spot fires and assess their progress (as well as any threat to populated areas), but obviously they could not match the mobility and range of aircraft. Suddenly, it was possible to catch a fire in its nascent stages or clearly see the path it was taking.

The involvement of aircraft in wildfire was a game changer. At the very least, it gave those on the ground a fighting chance against an enemy whose movement was often mysterious and unpredictable. It was only natural that as aviation technology improved, along with the skills of those who sat in the cockpit, aircraft would be utilized to even greater advantage. Innovation then, as now, was the key.

The early days of aerial firefighting were dominated by hard-headed, brave, and innovative pilots and engineers who were invariably toying with surplus World War II equipment to make an aircraft that could be tactically useful in fighting a fire. There is a legend in the industry that the first attempt at an aerial firefighter involved a beer keg strapped to a Ford Trimotor sometime in the 1930s. Other stories speak of a fifty-five-gallon fuel drum under the wing of a Stearman, emptied with the assistance of a pull cord leading to the cockpit. The veracity of these stories is long lost to the sands of time. But considering the history of our colorful industry, I would bet good money they are not only true, but also probably much scarier in reality than in rumor.

What can be ascertained is that some of the earliest documented attempts at creating certified and capable firefighting aircraft occurred at Nevada County Airport (KGOO) in Grass Valley, Northern California, south of Redding and west of Lake Tahoe. To this day, CAL FIRE and the USFS maintain an interagency Air Attack base at KGOO, which as of 2020 was staffed by an S2 Tracker Tanker and an OV10 Bronco Air Attack owned by CAL FIRE and operated by DynCorp, an Air Attack 17, and an AC690B Twin Commander owned and operated by Bridger Aerospace and manned by the legendary Todd White.

Todd, although unique in his own right and certainly far from typical in anyone's book, is a common type in the air attack world. He is a former smokejumper (smokejumpers are highly trained firefighters who provide an initial attack response on remote wildland fires; often they are inserted at the site of the fire by parachute—another practical application of aircraft in the struggle against wildfire) with decades of frontline firefighting experience in every climate from Alaska to Florida. In general, the best air tactical group supervisors (referred to as air attacks) are those that came from the elite firefighting communities of the hotshots or smokejumpers; they are former team leaders, division supervisors, or incident commanders who then take the path to air attack.

This career fork in the road takes members of a traditionally insular ground firefighting community and injects their experience into the world of wildfire aviation. From the outside looking in, wildland firefighting may seem like a monolithic community, much like how the military community is perceived, but the culture is far more complex. Although they face a common enemy and ideally serve a common cause, each different segment of the wildland fire management community grew from fundamentally different roots, resulting in unique organizational norms and, at times, diametrically opposed leadership priorities. It is this territoriality and tribalism—folks ostensibly united in purpose but sometimes speaking different languages—that makes aerial firefighting, already one of the most demanding jobs on the planet, even more difficult and dangerous.

Save for a few historic military engagements in the twentieth century, there is not a sustained aviation mission anywhere that comes close to encompassing the constant operational demands of weather, precision, and risk of aerial firefighting. When evaluated purely from the perspective of resources allocated to the issue and support provided, aerial firefighting may outweigh even the most demanding military aviation missions.

Take for example the aircraft themselves, the trusty steed upon which the pilot must ride to his destiny. Pick any aerial military engagement throughout history: the Battle of Midway, the Battle of Britain, MiG Alley. These engagements were waged in aircraft designed and engineered to fulfill the missions they were conducting, whether it was air-to-air combat, close-air support, surveillance, or long-range bombing. In almost every case, the aircraft involved were the result of years of complicated engineering, massive financial and governmental investment, and endless design refinements, all specifically intended to ensure they met the highest standards of military readiness.

These were the most sophisticated purpose-built aircraft, equipped with ejection seats, advanced sensors and technology, state of the art weaponry, thrust ratios that allowed them to climb almost vertically, and structural ruggedness that ensured survival in all but the most catastrophic circumstances. Combat pilots then (as now), squeezed into the cockpit knowing that while their jobs were undeniably and inherently dangerous, they would be conducted while encased in what were, at the time, the most advanced machines ever built.

Compare that to the scenario facing the typical aerial firefighter throughout history. While some of them may be former military pilots, they are often civilians coming from the far less glamorous—or even recognized—world of bush flying. These pilots transport cargo and small groups of passengers to mostly remote locales in places like Alaska, Australia, Africa, or the Canadian wilderness. They perform often mundane but crucial services such as surveys and supply regeneration to areas much of the world barely recognizes, routinely flying in and out of locations that most commercial pilots would consider too distant, inhospitable, marginalized, or just plain dangerous.

Rarely do aerial firefighters come from the commercial airline industry (although some certainly have), with its generous pay and benefits, predictable schedules, and perks that are too attractive to give up. It is the rare commercial pilot who wants to trade that security and comfort for the rigorous and far less well-compensated life of a bush pilot or aerial firefighter. These are uniquely adventurous men and women, drawn to bush flying for precisely the same reasons that most pilots run in the opposite direction: they want to test themselves in a challenging environment, they want to be of service, and they have a strong affinity for the natural world.

Bush pilots conduct missions that are, by all metrics, extremely hazardous. Not only because of where they are flying, but also because they fly in aircraft that are often converted from another purpose (that is, not "purpose-built") and frequently salvaged from the proverbial boneyard. It's not unusual for a bush pilot to find himself unwittingly at the controls of a plane long since retired by the third or fourth (or fifth) airline that owned it or a plane of dubious integrity acquired at auction from the government of an African or South American nation that could no longer afford to operate its cargo fleet.

There are mysteries involved in the world of bush flying, a great unknowable history that adds layer upon layer of complexity and risk to a job that is already among the most taxing in aviation. Bush pilots are resourceful, hardy men and women who must know not only how to operate a plane, but how to maintain and repair the aircraft as well, especially given the solo nature of the job and the wild, unforgiving locales in which they work. It would be grossly inaccurate, however, to suggest that bush pilots are reckless mavericks; the good ones, at least, are professional and careful because their clients depend on them for goods and services—and sometimes their lives.

And it is primarily from this spartan world, perhaps not surprisingly, that aerial firefighting draws many of its pilots.

"The truth is, aerial firefighting is a lot safer than it used to be," noted Jason Robinson, a former bush pilot who has been flying scoopers for Spokane, Washington–based Aero-Flite for many years and who helps trains other pilots on the CL-415. "But there's no room to be a cowboy when you're 150 feet above the trees. It is hazardous work and it's terribly unforgiving."

Mudslingers is at once the story of a former Navy SEAL officer who becomes immersed in the increasingly high-profile world of aerial firefighting and the story of aerial firefighting itself. The title is a reference to a retardant that is dropped out of firefighting aircraft—often called "mud," as the older type of borate–based retardant was thick and had a rusty brown color. Additionally, a lot of people who see any aircraft dropping any liquid on fire in North America often refer to it as "mud," even if it's water. As a result, those who drop mud, or "sling" mud, are sometimes called "mudslingers." It's a term that is said with affection and respect. Pejorative, perhaps, but also enjoyed by its recipients. As I found myself diving deeper and deeper into the world of aerial firefighting, I realized it is a story that so few Americans know. When thousands upon thousands of men and women risk their lives every year to protect them, it's only right that their story should be told.

In much the same way that *The Perfect Storm* folded the history of commercial fishing into the story of a single doomed vessel while also untangling complicated threads related to climatology and search-and-rescue missions, I hope to use my firsthand experience as a pilot and business owner in the field of aerial firefighting to chronicle both a personal journey and a historical narrative. This book is less the memoir of a Navy SEAL (there are enough of those already…) than the story of one person's love affair with flight and nature, and the strange confluence of events that led to a passionate and purpose–driven life beyond the military. Although, it should be noted that these two worlds will intersect in our story, as Bridger Aerospace was involved in assisting with the evacuation of friends and former colleagues during the U.S. military's chaotic withdrawal from Afghanistan in the late summer of 2021.

In telling this story, I want to take readers into the cockpit and into the lives of some of our crew members as they battle with the increasing scourge of wildfires and the danger that comes in unexpected places (I was personally involved in a crash that took the life of my instructor during a training session). These are tough but humble pilots, technicians, and mechanics in love with flight, the environment, and the intoxicating camaraderie of the fire base. They are committed to doing all they can to mitigate the existential threat of an ever-expanding wildfire season, and to finding meaning in a life of service. Risk notwithstanding—and make no mistake, aerial firefighting is extraordinarily dangerous work—most of them cannot imagine doing anything else.

"The best part about the job," observed Barrett Farrell, a chief pilot for Bridger Aerospace, "is knowing that you've helped someone. Maybe you've saved a life, maybe you've saved a house or a farm. Maybe you've saved some cattle or horses. Whatever it is, when you're flying home after doing thirty or forty drops in a day and you know the fire is smaller than it was when you started, that's a pretty good feeling."

It is a feeling and a lifestyle that cannot be replicated elsewhere, which is why they come back, year after year, season after unrelenting season, often leaving broken families and beleaguered spouses and partners behind.

"I don't know the exact statistics," said James Stewart, a Canadian former bush pilot who turned to firefighting more than a decade ago and who has flown Super Scoopers for Bridger for the past two years. "But I have to believe at least 90 percent of all pilots in this industry are divorced. It's just not a job that is conducive to long-term relationships."

Listening to Stewart—and nodding in agreement—was Pierre Dehaye, a wiry Parisian who formally trained as a chef in his teens and spent years working at restaurants and on cruise ships and for wealthy private clients before taking up flying as a hobby and eventually, in his thirties, making an abrupt career change. Like Stewart, he learned to fly scoopers in Canada (he now calls Montréal home); like Stewart, he is in his late forties and possessed of a calm demeanor that comes with proficiency and experience, which would seem to belie the profession's reputation for attracting risk-takers (a reputation that is unfounded, as it turns out, as the daredevils and cowboys are quickly weeded out); and, like Stewart, he is also divorced.

"It's difficult to be a partner when you are never home," observed Dehaye, who has been known to whip up impressively elaborate evening meals for the team after a long day of work in the field. "You just—" he paused, smiled—"drift apart."

Both men were sitting in the back of a scooper parked on the tarmac at Chico Municipal Airport, some seventy-five miles north of Sacramento, just beyond the Northwestern edge of the Sierras. Stewart and his fellow crew members had been deployed here for weeks in response to hot, windy, and arid conditions that have been known to turn brushfires into sprawling "events" in the region. The town of Paradise, which was nearly burned off the map three years ago, was only twenty miles away. Sixty miles to the east was Lake Tahoe, which would soon come under threat from the recently sparked Caldor Fire. At that moment, though, the Bridger team was on call, watching the weather and waiting for the phone to ring.

In fact, just a few hours later it did ring, but rather than being summoned to the Caldor Fire or some other new blaze in Northern California or Nevada, the team was instructed to pack up their gear and redeploy to Fox Airfield, located some four hundred miles to the south, in Lancaster, California. There, in the high desert of the Antelope Valley, at the northeastern corner of Los Angeles County, the waiting game would begin anew. The crew shrugged it off, like soldiers on deployment.

That is the job, after all. You never know where you're going to go or how long you will be there.

★ ★ ★

That assessment of aerial firefighting—pragmatic and respectful—is as true today as it was nearly a century ago. Modern weaponry at the firefighters' disposal is infinitely more advanced and effective, of course, but firefighters remain, at times, almost comically overmatched. There is nothing quite like a wildfire in a full-throated roar, ripping across tens of thousands of acres, to demonstrate the fearsome power of nature and its supremacy over all that man can throw in its path. The phrase "pissing on a bonfire" comes to mind.

"There's a question I get asked a lot by people who really don't know anything about wildfires," said Farrell. "'Why don't you just put the fire out?'" He smiled, repeated the question, with a slightly different emphasis, as if to reinforce its absurdity. "'Why don't you just put the fire *out*?' And I'll usually say something like, 'Well, it's not that simple.'"

★ TIM SHEEHY ★

Fighting wildfires will never be "simple"...or safe. But it has become smarter, and while it remains one of the most dangerous jobs—aerial firefighters face the omnipresent risk of crashing, and ground firefighters grapple not only with death or dismemberment in the heat of battle, but with long-term cardiovascular and pulmonary consequences stemming from smoke inhalation—it has become *safer*.

This is a good thing, because there was a time when it wasn't very safe at all. But with the growing threat of wildfires—which chew more acreage and displace more people from their homes with each successive year despite technological and meteorological advancements that presumably level the playing field—firefighting remains a job that is certainly not for the timid or the faint of heart.

For all of us in the wildfire world, external factors—particularly those on a vast scale like climate change—play only the slimmest of roles in our lives and work. It's not that we don't think about climate change; it's simply that fixation on the subject serves no purpose when you are in the heat of battle.

Wildfires have become a fixture of modern civilization. As they rage across hillsides, national parks, wine country, suburban neighborhoods, and even urban areas, there remains an array of opinions about the root cause of this persistent but evolving phenomenon. Some people are convinced that climate change is at the root of wildfire expansion and its increasingly ruthless scourge across North America, Europe, and Australia. To be sure, the damage caused to our environment by the industrialization of the twentieth century is no doubt a critical component of the growing frequency and severity of wildfires and needs to be addressed at multiple levels, from the policy-making hubs of Washington, D.C., and state capitols to the boardrooms of corporate America to the individual citizens whose collective carbon footprint creates enormous pressure on the environment. That said, any analysis is incomplete without the consideration of other aggravating factors.

Take for example, urban–wildland interface, as those in the industry call it. Essentially, this refers to the intensified interaction of wilderness with civilization: increased camping, vehicular traffic, and most notably, the construction of subdivisions in areas previously considered uninhabitable (and not without good reason). California is a prime example. With rolling foothills covered in sage and grasses, vast stretches of the state become tinder for fire when they are exposed to sun and wind (which is much of the year in the Golden State), dried out by weeks and months of arid conditions. When these hills, previously only ignited by a lightning event or a downed power line, are now subjected to thousands of citizens per day trafficking their area, the chances for incidental fire sparking are exponentially higher, as are the consequences.

"One of the biggest problems [in wildland firefighting]," said Joel Kerley, a career wildland firefighter for the Bureau of Indian Affairs, "is that we're not getting to fires soon enough or with enough resources to suppress or contain them. And that should be a mission that everyone involved in firefighting should want to achieve. There are people that will say, 'Well, you know, seventy years ago, we didn't always put them out.' And that's true. The Native Americans used to light fires on purpose because they understood the ecology of the time. But when you build mansions twenty miles out of Malibu to satisfy a market that is probably not going to slow down anytime soon, you then have an obligation to protect that market. It's a different equation today, and those people who want to revert to dealing with wildfires the way we did decades ago have the wrong objective in mind. It doesn't mean that we shouldn't be aware of that legacy, but we've got to figure out a way to get more fires out sooner by using all the different tools at our disposal."

★ ★ ★

On the morning of November 8, 2018, the residents of Northern California were on high alert for new starts. Those who live in typically fire-prone areas (presently, that means basically anywhere west of the Mississippi) are accustomed to checking the fire conditions daily. Certainly, this was true in the town of Paradise, which stood at the center of the Camp Fire, a tragic incident that constituted the single largest loss of life and property in the history of modern American wildfires. With an official death toll of eighty-six and an entire township virtually obliterated, the Camp Fire was the encapsulation of all the things that make twenty-first century fires uniquely challenging.

The residents of Paradise and the surrounding communities in Butte County were far from oblivious to the reality of their existence, which included the ever-present threat of wildfires. They had, for example, embraced proactive programs designed to manage vegetation and thus reduce the fuel available to wildfires as they approached populated areas. They had put into place an evacuation plan, which was triggered by an emergency notification system. But for all this supposed preparation and awareness, Paradise was ill-equipped to cope with the furor of the Camp Fire.

"We used to call Paradise 'The town that waits to die,'" noted Jim Barnes, a retired California firefighting pilot with some four decades of firefighting experience. "Well, one day it finally died. We always knew that that was going to be a major catastrophe there someday. One of the reasons is there's no escape routes. One road in, one road out. There was no chance for people to get out. You have to be able to shelter in place. I've given talks with fire chiefs, and we've addressed that issue. It's one of the most important things we could do—we have to identify and improve safety zones where people can shelter in place. Most of the people who lost their lives were trying to escape. But they had no way to get out. We're not talking about saving houses and cars; we're talking about saving lives."

A blaze sparked at 6:20 a.m. (likely from downed transmission lines belonging to Pacific Gas and Electric) in the sparsely populated town of Pulga, then traveled three miles to the west in an hour and a half, fueled by winds in excess of fifty miles per hour. According to a federal incident report released by CAL FIRE and the National Institute of Standards and Technology, by 8:30 a.m., the fire had already reached the outskirts of Paradise, with as many as thirty spot fires igniting around town. By 10:30, another three dozen spot fires had erupted in the heart of the town, including in and around subdivisions, trapping residents in their homes and cars. Within minutes, the town was completely overwhelmed as residents clogged the few roads that comprised evacuation routes. In addition to the loss of life, more than ten thousand homes and other buildings were incinerated, leaving most of the town in ruins. Total damages from the Camp Fire were estimated at more than $16 billion.

It was a fire of unimaginable destructive force. Except, of course, it wasn't, not to anyone who has ever done battle with wildfires or simply seen them up close, at their most intimidating. What the Camp Fire (and the concurrent Woolsey Fire in Southern California) really represented was the inevitable consequence of urban–wildland interface at its most destructive. Given the continued encroachment by civilization upon areas that were once wild, combined with climate change and other factors, it's fair to say that history will not view the Camp Fire as a freak event, but rather the first of such deadly megafires. And if we are going to continue to expand residency into areas previously deemed "wild," then we had best figure out how to both minimize the risks and cope effectively with the resulting wildfires with the least possible amount of human and property damage. Because they are coming. That much, at least, we do know.

But this book isn't about climate change, the causes of it, and how we need to stop it. It isn't about urban–wildland interface. At the risk of sounding melodramatic, when lives are on the line and we are in the air, climate change simply isn't my concern, nor is it the concern of any other aerial firefighter I have met. As is said by soldiers about war, "Ours is not to reason why, ours is but to do and die." Regardless of why fires are getting worse and all the international political maneuvering it will take to change the behavior of seven billion people, on a hot August afternoon in Montana, no one is talking about carbon emissions, green energy, or solar power when hundreds of homes are being evacuated. I can promise you that when the captain and first officer of a CL-415

Super Scooper are on final approach for a direct drop on the head of a fire, the effects of ocean currents aren't in consideration. And I can promise you that the rancher who is hurriedly evacuating its family and livestock isn't considering the impact of their cattle's flatulence on global warming.

By definition, an aerial firefighter isn't on a fire unless people and property are in danger or the fire is catastrophic, or both. So, when our mission is at hand, there is no time for global policy discussions. Our job is to fight the fire: stop it, put it out, or slow it down enough to make sure citizens can evacuate and ground firefighters can anchor a line. That's what this book is about: the history of those who fight fires from the air and how we can do it better in the future, as well as how I came to appreciate the enormity of that struggle, because no matter how you choose to look at the issue, the worst is yet to come.

I played with several titles for this book that I thought made sense for the aerial firefighting community. But unlike most glorified pilot groups, who fly high and fast, reach for the stars, blast through the atmosphere in sleek, shiny, superfast planes, the pilots who fight fire are a contradiction. They are pilots, yes. They love their planes and helicopters as much as other pilots do. They can be dashing, bold, and yes, even glamorous. But more often than not, if you ran into one on the street, you might mistake them for a fishing guide, or a rock climber, or a fish and wildlife officer, mechanic, cowboy, or truck driver. When they fly, they don't fly high and fast; they fly low and slow. The lower and slower, the tougher the job.

They fly in smoke, ash, debris, and usually in poor visibility and through tight terrain that would classify any other form of aviation operation "hazardous." Every year, pilots are killed or injured in the line of duty. Their mission is not breaking records or touching the stratosphere, and it is rarely recognized by the public. It is a much humbler task centered around saving homes, ranches, livestock—providing an escape route, protecting a piece of critical infrastructure, or helping a ground team hold the line against a raging inferno. As a result, I wanted a title that, while perhaps imperfect, highlights the people at the center of this amazing profession. The people that sacrifice their safety, their time with family, and in some cases, their lives, in service to others.

If you have ever seen firefighting birds in action through the sunlit smoke, or the crews at a bar after a long day, "Mudslingers" is about as close to the mark as you're going to get. It's a moniker that connotes a gritty no-nonsense outlook that is rooted in a mission to get the job done, and to go home—safe…alive—when that job is done. If ever there was a way that a nineteenth-century ship captain could be converted into a twenty-first century citizen, I believe it would be at the controls of a firefighting air tanker.

My hope is that you will take from this story a sense of how military aviation and civil aviation overlap and how together they have influenced aerial firefighting. You will learn about the tactics of fighting wildfires from the ground and the air. Most important, in that regard, is the indisputable fact that the primary role of aircraft in any conflict, really, is to support actions on the ground. Fire is no different. You will read about the amazing aircraft that have emerged in the somewhat quirky and niche world of aerial fighting. You will discover the agencies and governments that have been both pioneers and enemies of progress in fire aviation.

And you will learn about the early pioneers—hardscrabble hucksters, brave pilots, gritty ground pounders, ballsy smokejumpers, and myriad other characters who have been defined by fighting fire in the North American West. Canadian aerospace pioneers, giant seaplanes, forest empires, and aviation legends. Aerial firefighting has been a collecting point for fascinating people the world over. Heroism, luck (good and bad), greed, complacency, romance, skill, focus, recalcitrance, speed, bureaucracy, corruption, and swashbuckling are all words that frequently describe my daydreams and internal poetry about aerial firefighting.

I hope you enjoy the book, and most importantly, I hope it instills in you a deep respect and gratitude for the men and women who are fighting hard to protect our precious world.

CHAPTER 3

TALK-TALK

CENTRAL AFGHANISTAN
2012

When you get into the serpentine river valleys in the central part of the country, away from the mountains and the barren rocky landscapes, it's almost like a forest, like all the vegetation grows in one narrow strip of land. Counter to what you might expect of a notoriously arid region, parts of Afghanistan are thick and dense and lush. They call this area the Green Zone—not to be confused with the Green Zone in Iraq, where my previous deployments had been, which was the technical term for the secure allied zone in downtown Baghdad. This is different. This Green Zone is...well...green with foliage.

The first time I heard the term, I was confused because as folks continually referred to the Green Zone on my first Afghan deployment, I naturally assumed they were referring to a secure embassy compound in Kabul. In this case, it is simply a colloquial description for the jungle-like atmosphere that blankets a narrow pocket of land along the river—any river. Outside of that sliver of vegetation, there is only desert, a mountainous moonscape of rock and sand. But in the valleys—in the Green Zone—there is cover. And fighting through those valleys is slow and tedious and dangerous—creek to creek, tree to tree, hill to hill.

It was not what I expected, but by now, after three deployments with various SEAL units in and around the Middle East, I am accustomed to *it. I've also come to appreciate the value of teamwork and communication on these missions—not just between our SEAL units, which was usually no problem, but between all the allied units from other countries, host nation forces, and most importantly, the aircraft above us. Clearing a village can sometimes be a mundane task, with council elders agreeing to have lunch or tea and then willingly giving up an insurgent to avoid trouble or, more often, U.S. and Afghan forces carefully and meticulously cordoning off a village and preventing anyone from leaving while we knock on door after door in search of weapons or bad guys or some other target.*

And then there are the days when things get hot, with shots being fired before you even reach the village. So it is on this day, with a routine clearance quickly escalating into a firefight and the unearthing of IEDs and other evidence of a significant cache of weapons. One day becomes two, and two becomes three—a persistent multi-day battle as we slog through the valley, a hundred meters at a time. Thirty American troops and a hundred Afghan soldiers fighting through the population of an entire valley—giving ground, gaining ground—until we come across some Taliban in a bunker, maybe twenty yards away, dug in deep, almost Vietnam style, waiting until we're close enough before spraying us with belt-fed machine guns. A guy next to me, an Afghan national fighting on our side, gets hit in the face and falls. We all hit the ground and begin firing back. In this situation, there is little hope for air support because pilots are basically flying blind in the vegetation, unable to distinguish friend from foe. So, we ride it out, fighting for hours, picking up an inch at a time, trying to minimize casualties and make slow, steady progress. It's important to remember that the classic Navy SEAL mission—the kind the public thinks we do—is only part of what SEALs were tasked with during Operation Enduring Freedom (OEF) and Operation Iraqi Freedom (OIF). Although a small team of elite SEALs, Army Rangers, or other Special Operations soldiers would routinely go out as an exclusively American unit in the dead of night, just as often we would be tasked with these types of missions where we would be outnumbered ten to one by local troops who we were advising. The challenge was rarely our ability to fight; it was our ability to get the entire Afghan battalion to fight.

★ TIM SHEEHY ★

Finally on the third day, just as the canopy begins to break up, we get hit with a volley of RPGs and mortar shells. This is supposed to me my last mission in Afghanistan before transferring out on a South American narcotics rotation, and as the bullets and bombs blast all around us, I can't help but wonder whether I'll get out. I had already been wounded a handful of times on various other occasions, but there is nothing like one last mission to up the ante. It isn't panic I feel, so much as resignation. I'm only twenty-six years old, but I've already been at this long enough—have seen enough death and destruction—to know when the shit is getting deep. And right now, we're up to our knees—not up to our necks yet, but definitely our knees.

Fortunately, now that the Green Zone is thinning out, we have options, the most obvious of which is to call in an airstrike. We execute a nice flanking maneuver to neutralize the bunker shooters and grab some high ground, so we call a halt on the advance for a bit to evacuate casualties and get some air on scene. Coordinates are exchanged—friendlies here, bad guys over there—and before long, the Apache helicopters are whizzing overhead, launching Hellfire missiles as carefully as possible so as not to have any civilian collateral damage. It's quite a thing when the cavalry arrives like this. You feel a mix of relief and fear, knowing that your primary job in that moment is simply to stay put and hope the pilots—generally the best of the best—don't mistake you for the enemy. The last thing you want when an airstrike is called in is for a bunch of friendlies to start running around, instinctively looking for cover. When that happens, the pieces on the chessboard bleed into one—good guys and bad guys becoming equally likely targets. And whether it's friendly or enemy fire, the casualties need treating just the same. In this case, the biggest concern is Afghan partner forces getting out of position and getting hit by the air strike. So, you wait and watch, keeping the boys in sight, holding your breath as the missiles do their dirty work, cracking into the ground.

There are six men in my element, fanned out across a space of less than thirty meters, watching all of this unfold, when the air is pierced by a crack that I recognize—that we all recognize—as the sound of a Hellfire hitting the ground. I don't see it, but from the volume, I know it is close. And I know from experience that the sound emanates from an American weapon. The old Soviet bombs or Afghan IEDs? Those things explode with a dull thud and sometimes you'd get lucky. But an American Hellfire? It's an appropriate moniker—the concussive wave alone will knock you on your ass at bare minimum, even if you're not close enough to catch shrapnel. A bit closer and there's nothing to talk about. It's called brisance: the "quality" of an explosive material. High quality military explosives hit hard and have a concussive blast that will do serious damage. Homemade stuff or expired stuff will usually get the job done but with low brisance and a dull thud.

Time stops. The ground opens. A half dozen men are tossed about like ragdolls. Some time passes before I wake up. A few seconds? A minute? I don't know. I roll over from back to belly, shake off the cobwebs and the tinnitus. I had been hit by IEDs before, so I knew the drill, but this was far worse. Immediately, I hop on the radio and try to string together a few words.

"Cease fire, cease fire! Friendlies, friendlies, friendlies! Cease fire!"

Only then do I look around and conduct a quick head count. Four-five-six. Everyone is here, apparently alive. Struggling to their feet, woozily coming to terms with what has happened, reality sinking in the way it might for a boxer who has just been knocked out. Between us, almost impossibly, is a hole in the ground where the Hellfire landed, a gaping gash of ripped rock and sand. It seems large enough to swallow us all at once. That no one was killed or maimed is a stroke of good fortune beyond comprehension. And yet, *I can't help but think,* why did this happen? What went wrong?

Well, it was pretty obvious what went wrong. The one *thing we all train to prevent in calling in airstrikes: the friendly aircraft shooting at the good guys instead of the bad guys. It's shockingly common in war despite years of training and policy to stop it. When a ground team calls in an airstrike, they give the friendly position* and *the enemy position so that the air crew can ensure whatever munitions they use won't also hurt the friendlies. Unfortunately, sometimes the ground team mixes them up or the pilot flips them around in his head. Either way, it's a tale as old as war. Since the beginning of indirect fire, artillery, naval gunfire support, and then air support— friendly munitions killing friendly troops has been one of the biggest killers of troops on all sides in history. During the first Gulf War, we beat Saddam in one hundred hours. Easy day. But in the end, we killed more of our own people than the Iraqis did. In World War II, at Guadalcanal, the U.S. Navy sunk at least two of its own ships with friendly fire as the naval community grappled with the realities of complex night engagements. Obviously, in this instance, something had gone awry with air attack-ground integration. And the scary part was that we were among the best in the world at it. We were the best on the ground and the best in the air, and it could still get all messed up and cost people their lives.*

There had to be a better way.

★ ★ ★

I wanted to fly. It was that simple. I grew up near a lake in a semi-rural part of Minnesota, where sports and outdoor activities are embedded in your DNA, and I was involved in just about everything you would expect of a kid from that region: fishing, hunting, hockey, treehouses. But at heart, I was a Huck Finn type of kid—throw the door open and let me go outside and dig tunnels in the dirt or the snow or build forts out of deadfall. The outdoors was a vast, mystifying playground, and I couldn't get enough of it.

On some level, even then, I also understood that there was a responsibility to the land, that it wasn't just there for the taking or exploiting, that it was meant to be appreciated and protected. I had no formal training in this regard; we weren't environmentalists. My father, Richard Sheehy, had worked in construction when he was younger, then moved into real estate management and investment before an accident left him in a coma when I was just a little boy. His recovery was slow and incomplete, and he never went back to work full-time after that, but he remained a vital and important presence in my life, always visible, supportive, deeply involved, and a great teacher and mentor. His accident, although ending his own career before it had peaked, helped my brother and I reach incredible heights in our lives because we got to spend a lot more time with the best kind of mentor a young boy can have—his dad! My mother, Denise, was a stay-at-home mom who was the consummate midwestern Scandinavian mother. She took great care of us, kept us on the straight and narrow, and made sure all our crazy activities were supported. I have a brother, six years older, who was, along with my parents, the best teacher and supporter I could ever dream of. I owe almost all of my business success to my brother, Matt, and have learned countless lessons from him since I was a baby. I didn't have a broken home or hardscrabble upbringing; this is not a rags-to-riches story, and I won't try to make it such. Ours was a happy, well-adjusted family in which hard work was expected and adventurous behavior encouraged.

Within reason, of course.

Some parents might have worried that their eight-year-old son was taking informal flying lessons from a neighbor, but not mine. Like I said, we lived near a lake, and in the winter (which in Minnesota stretches from roughly late October until the end of April), the lake would freeze over, and our neighbor would land his little single-engine Piper on the ice in a display of casual aeronautics that I found mesmerizing. Harry Thibault lived with his wife Molly just to the north of us. They were a big Catholic family with several grown children and were always showing me the example of service in life, going on mission trips and bringing in exchange families from all over the world who needed specialized medical treatment. He was a natural storyteller whose gentle demeanor belied a streak of toughness. Harry was a Navy veteran who had flown during the Korean War, and it's no surprise that his son Steve dreamed of flying F-14s when he was growing up. Some minor issues with vision disqualified him from such an elite endeavor but did not prevent him from acquiring a civilian pilot's license, and he eventually became a certified instructor.

By the time my father got married and had children, he and his parents' family had become estranged, making cohesive family gatherings rare, and we had little to no relationship with my grandparents on that side. My mom's dad had died when she was young and her mother had died when I was young, so I didn't have any grandparents to speak of. Harry and Molly, nearly a generation older than my mom and dad, became my de facto grandparents, and Steve an uncle of sorts. I don't recall ever being scared in that Piper, not even the first time Steve took me up in the air. As the plane lifted off the ice and climbed above the Minnesota whiteness, I could feel my heart beating faster, then settling into a peaceful rhythm. *This*, I thought, *is where I belong.* By the time I was ten years old, Steve was letting me take the controls once in a while. By age eleven, he was taking me up in a single-engine Cessna and giving me lessons. For a few years, I was the rare kid who knew how to fly a plane but had never been behind the wheel of a car! I received both my driver's license and student pilot's license at the age of sixteen. But even the freedom that comes with being able to drive on your own—the dream for many American teenagers—could not compare to the raw exhilaration of flight.

I was a solid student and a good, versatile athlete who managed to avoid trouble. At various times during my adolescence, I excelled at hockey and swimming. I enjoyed sports, and they generally came easy to me, but nothing compared to the thrill of flying, and I became obsessed with the possibility of becoming a fighter pilot. I watched *Top Gun* over and over, and I imagined myself as a real-life Maverick. When I was little, I dreamed of becoming an astronaut, and as I got older, in high school, I did more than just dream.

In a fit of nerdy ambition, I became a frequent participant at NASA's Space Camp program in Huntsville, Alabama, where high school students are immersed in astronaut training and techniques, including the use of equipment adapted from the actual NASA astronaut training program. But Space Camp is much more than an amusement park experience. It is a challenging academic experience in a highly competitive environment, at the end of which winners are determined and awards are presented for leadership and teamwork, among other categories. Some kids attend Space Camp on a lark. They tend to be both humbled and miserable. But those who are into the experience and are interested in careers in the aerospace industry—particularly those who believe they have the *right stuff* and want to enter the insanely narrow funnel that leads to becoming an astronaut—find Space Camp to be an immensely rewarding experience.

I fell into the latter category. I took it as seriously, if not more seriously, than any hockey game or swimming meet. Somewhere in my parents' house back in Minnesota are roughly a dozen little trophies and awards that I accrued at Space Camp. Not that hardware is important, but it's an indication of how much it meant to me. I really loved it, and I treated it like it wasn't just a vacation. I studied the manuals and prepared for the simulations. I *practiced*. I sincerely wanted to do well, which was not the case with most kids at Space Camp. And you know what? I still benefit from that experience. I was an Army Ranger and a Navy SEAL; I've been a firefighting pilot and a mini-submarine driver, and throughout all of it, the stuff I learned at Space Camp comes in handy to this day. In America today, more kids aspire to be social media influencers than astronauts. Programs like Space Camp can help to reignite that passion in our youth.

As we look at other nations (China, most notably) that are superseding us in the journey for space and military dominance, we must encourage our youth to engage. America has served as the beacon of freedom and opportunity for two hundred years; we must fight to keep our edge. People are not drowning in boats, climbing fences, or falling from the landing gear of airplanes while trying to emigrate to Russia or China, but they take those risks to come here. It's our obligation as Americans to protect our future.

★ ★ ★

My parents were not opposed to my entering the military, but they did insist upon a reasoned discussion of goals and motives. As children of the Vietnam era, they had come of age during a time of immense political turbulence surrounding an unpopular war fought mostly by soldiers who had not volunteered for the job. By the time I was in high school, the draft had been eliminated, and the military had long since transitioned to a highly trained unit of professional soldiers. In the wake of 9/11, the country was awash in patriotism (and a strong desire for retribution). This made the notion of a career in the military easier to accept for parents who might otherwise have feared for their children's lives and questioned their motives.

Like everyone in my generation—like those in the generation before us who were alive on the day of the JFK assassination—I remember exactly where I was when 9/11 happened. I was a sophomore in high school and was in history class when a ruckus in the hallway got everyone's attention. We all just kind of naturally drifted downstairs to a classroom with a TV in it. As a bunch of fifteen-year-old kids, we were talking to each other like we knew what was going on.

"It's probably a gas line explosion or something. No way an airplane would hit a building in broad daylight, it's clearly a mistake by the news"

"It was probably a helicopter not a plane…"

"Maybe it was a crazy guy who did it on purpose?"

None of us realized what a pivotal moment in history we were watching unfold; it was just a normal Tuesday morning, and frankly, as bad as we felt, there was also the underlying feeling that we were thankful to have something change our schedule for the day. Tragic, yes. Terrible, yes. But as is typical with teenagers, this was a reason to gossip and get out of class for a bit.

We chatted, feigned serious consideration, and discussed with imitated wisdom what must be going on. Then it happened.

We were watching the news as the second plane hit. I will never forget the feeling in that basement classroom. No one said a word. It was complete and utter disbelief. We may have been dumb fifteen-year-olds, but we were smart enough to know that two commercial airliners don't hit skyscrapers by accident. It was at that moment we all knew that our lives would change forever. Some more than others. The America of 2001 was not the same as 1941, nor was the conflict. Most of these kids would go on to liberal arts colleges and spend the rest of their adult years in the comfort of a country they could criticize freely. For a small number of us, though, it meant we would be spending most of our young adulthood fighting our nation's longest war.

My parents were supportive but challenged me. "Why do you want to do this?" they would ask. "What do you think it will be like? Do you know what you are getting into?"

Like a lot of young men in the early 2000s, I was eager to fight. Who needed college when the world seemingly was on fire? I was not a disrespectful kid, and my parents were not authoritarian. But there were some tense moments in my household in those days, as we sparred over what was right and what was wrong. In the end, we reached a compromise of sorts—an agreement on what path seemed to make the most sense. I was a good student, and my parents naturally wanted me to go to college before embarking on a career in the military. The best way to combine those two ambitions was either through an ROTC program, or, better yet, by applying for admission to one of the service academies.

Admission to any of the service academies is highly competitive, but I had a strong academic and athletic record plus a solid resume of volunteer work and extracurricular activities. I was the captain of three sports teams, got good grades, and had strong SAT scores. Combined with enthusiastic recommendations from mentors, this resulted in my acceptance to the United States Naval Academy in Annapolis, Maryland (USNA). To be honest, receiving the acceptance letter didn't provoke the flood of pride and excitement that typically comes with acceptance to one's college of choice, but rather left me with a feeling of profound ambivalence. I was excited, but also conflicted. I wouldn't be on the battlefield soon.

Hey, I got in. That's great...I guess.

My parents were thrilled, in part because it was a significant achievement (and who doesn't want to see their child succeed?), but also because it delayed the possibility that I would be on the ground in Iraq or Afghanistan, at risk of getting ripped apart by an IED or a sniper's bullet. For a few years, at least, I would be safely tucked away in Annapolis. And by the time I got out? Well, who knew? Maybe the war would be over.

Maybe not.

Like any other massive bureaucratic organization, the military allows only so much to be understood from the outside looking in. But once I was at the Academy, I began to see that we were in a challenging situation. We were fighting two wars at once, and while most people didn't realize it at the time, neither one of them was going very well. In my plebe year, there was a briefing for the midshipmen from the commandant where the message was basically, "We're taking a lot of casualties over there, the face of this war is changing, and we need more Marines, we need more of you on the ground." And I could see then that becoming a fighter pilot was a selfish goal. It wasn't what the Navy needed from me, and it wasn't what the country needed. After having been at the academy a few months and seeing and meeting several officers from across the Navy, I realized the Navy SEAL teams or the Marines were where I belonged in this new warfare environment. A lot had changed from our perception of fast jets and submarines in 2000 to the reality of stone age era enemies in the desert in 2004.

Flush with youth and patriotism, my first inclination upon receiving this news was to resign my commission from the Naval Academy and immediately get in the thick of the fight as an enlisted man—not much different than it had been a couple years earlier when I was still in high school.

"Sir, I can't just sit here and watch the war being fought anymore," I told my company commander, an Academy upperclassman named Keith McGilvray. "I want to be in the fight. Now."

He didn't put up much of an argument, basically just said, "Okay, well, draft a letter of resignation and we'll start the process." A pause. "You're sure about this?"

"Yes, sir."

McGilvray was a no-nonsense upperclassman who ended up a decorated Marine Special Operations Officer. He had put me through plebe summer boot camp, so he knew me from the beginning. He was also the captain of the hockey team at USNA.

"I'll route this up the chain for you, but you need to have a conversation with someone in the meantime. He'll meet you at Dahlgren tomorrow night. Look for Mr. McDonnell."

Others, apparently, were more determined to save me from my own misguided sense of honor. Dahlgren Hall was an old drydock building that had been converted into an ice rink and an aptly named restaurant called Drydock that was off limits to most midshipmen. As a plebe, I was on the lowest rung of the Academy ladder, but I had been instructed to meet someone there, so I followed orders, albeit uncomfortably. I knew only that his name was McDonell and he would be senior to me. I wandered around the restaurant, awkwardly scanning the obligatory name tags of various folks, until I finally found him sitting alone in a booth. A cursory glance at the hardware on his uniform revealed him to be a full-bird Army Special Forces colonel. And there I was—a Naval Academy plebe. I was nothing. I was lower than nothing compared to this guy. I had no idea why an Army colonel wanted to talk to me.

"Excuse me…sir?" I said, as I waited for an invitation to join him.

He looked up. "You're Tim Sheehy?"

"Yes, sir."

"Have a seat, son."

He went on to explain that he had recently returned from an eighteen-month tour in Iraq and was now stationed in Washington, D.C. It wasn't uncommon for officers from various branches of the military to be assigned to the Academy as mentors while in D.C., and Colonel McDonnell had been attached to the hockey team as an officer representative (every extracurricular organization at the Academy was required to have an officer assigned to ensure that the club, team, or organization was providing training and guidance consistent with the goals and expectation of the Academy, such as officer development). My company commander, as captain of the hockey team, had mentioned to the Colonel that he had a boneheaded young plebe who needed some sense talked into him. Colonel McDonnell had been involved in Special Forces and had been made aware of my desire to become a Navy SEAL, and so it was decided that he should have a sober conversation with an enthusiastic young plebe who seemed determined to throw his career away.

"I heard you want to quit the Academy and go fight the war."

"Yes, sir, that's the plan."

"May I ask why?"

It seemed a question with an obvious answer, but I explained my rationale, my frustration over being on the sidelines while so many people were fighting and dying. Like many idealistic young kids in a thousand wars before mine, I was worried the war would be over before I could do my bit to help.

"That's a noble perspective," he said. "But I have to tell you, son—I just came back from a year and a half in Iraq. These wars will last another decade or more. You don't have to worry about missing anything. Furthermore, we have plenty of door kickers and shooters. What we need now, with this counter-insurgency movement, are leaders who can think on their feet and make the right decision. If you really want to be of service, that's the best way to do it. More than any other war America has fought, this is the sergeant's war. The captain's war. It's not a general's war. We need sharp young leaders who can lead their men in the field and bring them home. If you really want to do what's best for the country, finish your training and join us as a *leader*."

As an incentive, he recommended a new program that would put certain promising midshipmen at the Academy on an accelerated program designed to facilitate integration between various branches of the U.S. Special Operations Forces—specifically, between the elite Navy SEALs and the Army officers (Rangers and Green Berets) who they frequently answered to while on deployment. In almost every deployed environment during the global war on terrorism, SEAL units are assigned, at some level, to an Army command, and therefore always work for the Army. As you can imagine, if you work for someone, you'd better figure out how to speak their language.

In every other special operations force, you have to do time in the regular military before you become eligible for Special Operations Forces (SOF) duty. Ranger, Marine Forces Special Operations Command (MARSOC), Delta, and Green Beret officers all serve time in the regular line before they can compete for SOF spots. The SEALs are unique, especially for graduates of the Naval Academy—you basically pin on your officer bars in Annapolis and go straight to BUDs (the rigorous SEAL training program that weeds out 80 to 90 percent of all applicants) and become a SEAL. So, there was a problem with SEAL teams integrating with Army commands—there was a language breakdown. That's really all it was. But it needed to be addressed.

The way SEALs plan a mission, I discovered, is different from the way missions are planned in Army Special Forces. The way we rig our gear, organize our units, build our command structures—all different. Neither is right or wrong—just a different approach. Eight guys swimming to a beach at night to emplace explosives have very different operational needs than thirty guys patrolling through the desert. The communities were different because their roots were from fundamentally different origins. But when working together, you must figure out how to speak the same language. It's like taking a star soccer player and putting him on a football field and expecting him to just blend right in. He might be a great athlete, but until he understands the game and what is expected of him, he's not going to be very effective. So, some enterprising Special Operations Command (SOCOM) folks decided to address the issue by training people in leadership positions about how the Army works. For a while, they were taking SEAL officers and sending them to Ranger School, but that's like taking a college grad and sending him back to high school, risking injury and chewing up valuable time. Then they decided to start earlier, taking SEAL students between phases of training. The problem there, however, was that Ranger training is so rigorous that candidates were getting beat up and were forced to delay entering BUDs. Fine-tuning the schedule was a challenge.

In the summer of 2005, after my plebe year, I ended up being a test case—the first midshipman to be sent off to Ranger indoctrination at Fort Benning, and subsequently Ranger School, Airborne School, and other training, while still a student at the U.S. Naval Academy. I was fortunate enough to follow these schools with temporary duty rotations with Army Special Forces units. While in these schools as a young naval midshipman, I was surrounded by professional soldiers, most of whom were several years older than me. I put my head down and went to work. I understood where I was. A lot of these guys were already soldiers with impressive résumés. Guys who had been on deployments and seen real combat duty. One was a Marine recon guy; another was a Green Beret going through Ranger School. I was a nineteen-year-old college student. But I learned quickly that working hard, being competent, and not behaving like an asshole goes a long way. It's not that complicated. I also learned that leadership comes in many forms. The traditional naval leadership lessons, although tried and true, wouldn't always apply to leading men on the ground. In fact, to this day, one of the best leaders I ever saw was a young army corporal Ranger student, Matthew Thunderhawk. He didn't say much and didn't outrank anyone, but he always led by example and showed me that even without the authority to lead, one can still be a leader.

Throughout this unique journey, I would meet several Army counterparts who would become lifelong friends and provoke conflict within me in regard to where my loyalties resided: Army or Navy! Lee Dingman was an Army Ranger officer at the Naval Academy while I was there and was also a mentee of Kevin McDonell. He became my case handler for my new Army program, and he and I became close friends who would go on to found AVT together and have many other life adventures along the way.

By the time I graduated from the Naval Academy in 2008, I was a commissioned officer in the U.S. Navy, as well as a U.S. Army Airborne

Ranger. Eighteen months later, I was a Navy SEAL officer who, two weeks after SEAL graduation, was on his way to Iraq. Although I had deployed before as part of my Army training, this was my first deployment as an "operator," and I was very excited. I was assigned to a joint task force in Iraq in the winter of 2009 working on a joint operation with an Army Special Operations squadron. As soon as I got there, I met up with some of the guys from my Ranger days, which immediately showed me the value of the cross-training program I'd gone through.

Through multiple deployments with Special Operations over the next five years, from Afghanistan to South America, I became fascinated with the process of optimizing our effects on target through enhanced aerial asset integration. Clear, concise, and prompt communication between various constituencies is essential—especially between ground forces and tactical air support. In some cases, the synergy worked impressively; other times, not so much. But the longer I served and the more I was exposed to complex military engagements, the more I came to appreciate the value of communication and shared resources. Territoriality struck me as perhaps the biggest obstacle to achieving successful outcomes in the military, and I was excited to be part of operations that required the sharing of resources and information. But it was often a struggle.

In the fall of 2010, I was part of a SEAL unit involved in the search and attempted rescue of a British aid worker named Linda Norgrove, who was kidnapped and taken hostage along with three Afghan colleagues by Taliban forces in the Nangarhar and Kunar Province. Norgrove was working in Afghanistan as a regional director for a company called Development Alternatives Incorporated, which had contracts with various government agencies in the U.S. and other countries. Rumors persist to this day that Norgrove was also working undercover for the British intelligence agency MI6, but they have never been substantiated and were vehemently denied by her family. Regardless, she was a Western hostage now being used as a political pawn by the Taliban.

★ TIM SHEEHY ★

At the time of the kidnapping, I was assigned to a classified joint task force and, once again, was acting as a liaison between "Blue" and our Army counterparts who, at this time, were the Rangers. I enjoyed the assignment as I once again was able to meet up with a lot of my former Army colleagues. It's amazing how small the world of Special Operations is. And since Linda was a U.K. national, it would have technically been the right of the British Special Air Service (SAS), arguably and traditionally regarded as the finest commando unit in the world, to lead this mission. But given that SEALs had been operating nonstop in the mountains of Northeastern Afghanistan for almost a decade, we were best positioned to action this objective. During the Iraq build up, our British and Army colleagues largely moved their focus there and had moved out of this area of operation (AO) for many years by this point. This operation was in the toughest part of Northeastern Afghanistan, an area that included Tora Bora and the Korangal Valley—a.k.a. the Valley of Death—a region that had already been home to some of the bloodiest battles in the Afghanistan conflict, including the Battle of Kamdesh in 2009. As is always the case in hostage rescue operations, time was of the essence. The longer a hostage is missing, the less likely it is that they will be recovered—or at least recovered alive. But first and foremost, we had to locate her, which proved to be no easy task.

For the better part of two weeks, we saturated the region with specialized collection aircraft absorbing all manner of data available from signals intelligence to imagery intelligence in an effort to locate the hostages; we compiled a mountain of sensor data. I was the liaison between the air task force and the 101st Airborne, which was the battle space owner at the time. We were pushing tons of patrols out into the valley, walking around and clearing building after building, chasing every lead, trying to find her, monitoring lines and picking up radio chatter—listening for any sort of clue.

It became a chess game played against the backdrop of a ticking clock. When you're a shooter with special operations working on a hostage rescue mission, your job is straightforward. You enter a room, clear

★ MUDSLINGERS ★

the room, and shoot bad guys before they can harm the hostage. It's not an easy job by any stretch of the imagination, but it's clearly defined. When you're a commanding officer, intelligence analyst, or team leader, the job is different. It's a tactical role, one in which extremely difficult decisions must be made in an atmosphere that has changed dramatically over the years. You have to understand that the whole hostage rescue mindset was developed in the 1970s and 1980s when every hostage situation was virtually the same. A bunch of extremists would grab some hostages on a plane or in a building, and everyone would know where they were, and then the extremists would make a bunch of political demands associated with the fact that they had hostages. Bad guys inside with hostages, good guys outside waiting for the opportune moment to execute a daring rescue or create a negotiated settlement (think the Iranian Embassy siege, the MS *Achille Lauro* hijacking, the Waco siege, or dozens of other examples from that era).

The whole paradigm of hostage rescue was developed based on that mindset. But as we learned in the post-9/11 world, the modern extremist approach to taking hostages was markedly different. In general, in the Middle East in the twenty-first century, extremists weren't taking hostages for political demands; they were taking them to make a statement. They were taking hostages so that they could be murdered on camera in terrible ways to spread fear and intimidation, or they held hostages for long-term political leverage, not short-term demands. Long-term leverage could be exercised via increased credibility within the competitive world of terrorist networks and, occasionally, by getting a concession from the U.S. or coalition governments. We would never know where the hostages were; their first tactical move was to move them and then hand them off to another group, who would move them again and again and again. Knowing our capacity for rescue and targeting, the idea was to prevent us from knowing the location of the hostage until they were brutally executed. In that new reality, what exactly were you negotiating? What leverage did you have? The answer, of course, is none. So, the whole arc of hostage negotiation changed from one of negotiation, patience, and timing to, quite simply, speed and tactics. Find the hostages as quickly as possible, get them out immediately.

The three Afghan colleagues were released on October 3. Over the course of the next week, concerns mounted that Norgrove would not be released but more likely would be sold up the terrorist food chain, moved across the border to Pakistan, and then the worst could happen. That's one of the things that was so confounding about the twenty-year war against terrorism—the fact that we never could figure out exactly who the enemy really was. Certainly, it wasn't some monolithic entity. Half the time you didn't even know who you were dealing with—the Taliban, Al Qaeda, the Haqqani network, Iran, Quds Force, ISIS. Each had its own tactics and goals, and to conflate them was to risk completely misunderstanding the nature of a mission and risking the life of a hostage.

In the end, we were able to locate Norgrove, but the process was maddening and the final result dispiriting, as she died from wounds suffered during a pre-dawn raid on the kidnappers' compound. We had the greatest technology on the planet—the best intel, the most powerful weapons, the most talented pilots and soldiers—and yet it still seemed like there was so much we were trying to figure out on the fly. I couldn't help but think that aerial–ground coordination and optimization of capabilities was a long way from where it should have been.

At that time, I was still young and optimistic enough to believe I would be a career military man, but part of me wondered about the possibility of someday starting a business, one devoted to solving these types of problems—enhancing communication between constituencies that might not ordinarily play together so well. I had no idea what it might look like, but it was a seed. Planted, if not immediately nurtured.

I actually remember the phone call like it was yesterday. A week or so after this mission, I was passing through an airbase and had the rare opportunity to make a phone call home (this was before Wi-Fi calling and cell phones that could easily call back to the U.S.). I usually called my parents or my future wife. This time I called my brother, who was living in New York at the time and building his business career. After the usual catch-up chatter, we talked about what I had seen.

"I don't know anything about business, Matt, but I am seeing some interesting stuff over here regarding aerial technology. It's in its nascency, but how we can gather information from aircraft and share it with the ground in real time? If someone had the chops to do it, there could be a lot of business potential there."

"What kind of business potential do you think? Military sales or other type of stuff?"

"To be honest, I don't know, but my sense would be to refine some of what we are doing and commercialize it for other stuff. I have no idea how to do it, but you should keep your eye on it."

"Well, you don't know anything about business, and I don't know anything about military aviation. Maybe someday we can put it together and make something happen."

"Yeah, someday, maybe. Anyway, I'm safe and sound over here, should be on my way home in a few weeks. See you soon!"

As it happens, I would see him shortly after I got home from that deployment and just happened to have dinner with someone who would eventually be our main investor in our business. Small world indeed.

Two years after the Norgrove mission, a Hellfire missile ripped through my unit when we were mistaken for enemy troops. While we were all lucky to escape serious injury, I was struck again by the importance of communicating with others and understanding the nature of a given problem. It was vital to encourage people out their silos and into more transparently cooperative and mutually beneficial relationships to work for the collective good.

★ ★ ★

One of the things I cherish about my military experience is that, while it was relatively brief (eleven and a half years, four of which were my Academy and Army exchange time), I got to experience so many different things. I worked with multiple Ranger battalions; I deployed with various SEAL teams, several Special Forces Groups, law enforcement; I worked on an undersea team in Hawaii with multiple intelligence agencies, and I worked on the ground with conventional military units in Iraq and Afghanistan. I participated in operations on four continents, and I got the opportunity to deploy with a broad spectrum of military and intelligence professionals, not to mention local militias and commando units from nations like Iraq, Brazil, Afghanistan, Australia, Lithuania, and more. The diversity of my experiences gave me an interesting perspective on things. People are inherently tribal by nature, and this is especially true in warrior culture. Everything tends to be viewed through the prism of the tribe, the regiment, the team, the squadron. This is good, as it creates bonds of loyalty as well as team cohesion, which leads to operational excellence. But it also can create steep barriers—barriers that can ultimately be fatal. As I would later learn in aerial firefighting, that trend is universal.

I was fortunate enough to bounce around the theater for my career, so instead of looking at an issue from one angle, I was turning it upside down and inside out and looking at all the different sides. Sometimes this made me unpopular with traditionalists, both while I was in the military and after I got out. By the time I left Afghanistan, it had become clear to me that our counterinsurgency efforts were failing primarily because we were too addicted to hunting an ever-changing roster of terrorists. The stability campaign was always subjugated to the direct action campaign. In true counterinsurgency, the goal is to build local capacity—to create stability through a whole-government approach, so that over the long term, a stable society takes the place of chaos and repels extremist elements. This takes time and effort and diligence. It takes patience, maturity, and a long-term view, all things America has very little of in the twenty-first century.

Many people in the military—Special Forces in particular—simply wanted to go out every night and find ways to kill as many bad guys as possible. And it obviously wasn't working. We had been running around killing bad guys for a decade and we were not making any progress. We needed a different approach, one that included a concerted effort to improve stability through infrastructure, economics, and education. It takes time. You can't just walk in, destroy a country, and then leave and expect everything to be okay. We invaded Germany and Japan in 1945 and we are still there today with tens of thousands of troops. We're still in the Philippines. We're still in Guam, Cuba, South Korea, Okinawa, Kuwait, Bosnia. Because that's how long it takes to provide stability after you destabilize a society. In Afghanistan, for some reason, we labored under the illusion that we were going to come in and bomb the shit out of the country, kill a bunch of bad guys, and then leave a couple years later…or ten years later…and everything would be fine. It doesn't work that way, and never has worked that way. It can take days to win the fight; it takes decades to win the peace.

This perspective, founded on the concept of building coalitions, helped me immensely in developing a successful aviation company (two aviation companies, actually, one of which, Ascent Vision Technology, develops and builds drone technology for military intelligence, surveillance, and reconnaissance and was sold for $350 million in 2020) in a relatively short period of time, despite working in a field that is not only highly competitive (with lives at stake), but complicated by layers of governmental oversight and territoriality.

I hosted an event not long ago with regional business leaders and they all wanted to know, "How do you go from a tiny company to the size you are now, and how have you done that with no business experience?" You're a legal adult at eighteen, but the brain continues to develop into your late twenties. I spent my entire twenties fighting a decentralized conflict against a fragmented and diffuse enemy that would adapt and change every day. It shaped my brain and my thinking to be constantly adaptable. Instead of crafting and adhering to a rigid strategy, I would determine principles and priorities that could guide my team's actions a thousand times a day. I was fighting an enemy that would change tactics every day, change methods every day. What I realized was that we were trying to fight an enemy that was playing golf while we were playing football. We were running around the golf course in football pads trying to tackle an enemy that was hitting a drive off the tee and then walking into the woods. It didn't work.

We spent twenty years playing whack-a-mole in Afghanistan. We'd kill one guy, a cell leader, supposedly important, and what difference did it make? None. The next day he'd be replaced by a younger, more ruthless, more aggressive guy trying to prove himself to all the other insurgents. We killed so many people, so fast, and they were always immediately replaced by someone angrier and more radical. There was no way to keep up, no way to shoot or bomb our way out of the mess. Eventually we learned that the only way to make some progress was to form coalitions, work with the locals, and try to build trust. It was exceedingly complicated, time-consuming work, and it required thinking outside the box—not always a strong suit of the military. The most important thing—in a way, like parenting—was just to be present. Be there to let the institutions develop to a point where they were strong enough to stand on their own.

★ ★ ★

The end, for me, came in 2014. I had already sustained several injuries of varying degrees of severity—concussions, bullet wounds, shrapnel (some reported, some not, because, after all, if you report an injury, the only thing you know for sure is that you're coming out of the fight indefinitely). The only time my body was actually hit by a bullet, oddly enough, was not a knock-down, drag-out gunfight, but a very lame night engagement in Afghanistan. It's not even worth a story, it was a routine "walkabout" where we had to move to an adjacent observation point to do a resupply of the Afghan troops manning it. As we were on our way, we took some potshots from the ridgeline, sent a few rounds back their way, took a few more potshots, and then moved on. Compared to many of the firefights I had been in, this was literally "nothing." But it still got the blood pumping. We dropped the supplies off, and as my body calmed down from the climb and the adrenaline of the gunfight, I noticed my arm hurt. I thought maybe I had hit it on a rock when we were maneuvering during the firefight. By the time I got back to base, it was pretty sore. I learned later there was a bullet in my arm! It's still there today by the way. My various IED exposures had produced a couple traumatic brain injuries (TBIs), some shrapnel hitting my right leg and head. But overall, I was relatively unscathed. I had never reported any of these injuries because I didn't want to be taken away from the most cherished time in a SEAL officer's career: my time as a platoon commander and team leader, when I was actually leading my team on missions. After that period of your career, the future is paperwork and politics. So, I thought (as did my wife) that after all that, I was in the clear.

We were wrong. The most impactful injury of all was still to come, on a dive training exercise in Hawaii.

Around 2012, as the drawdown in Afghanistan became apparent, there was a movement within the military, and within Special Operations specifically, to strengthen our maritime core, particularly in the Pacific Theater. In the previous decade, with ceaseless fighting in Iraq and Afghanistan, emphasis had naturally been on developing and training ground and aerial forces capable of executing the war on terrorism. Of all the SEAL teams, the smallest and most specialized is the Seal Delivery Vehicle (SDV) team, but with interest in maritime capability surging, there was active recruitment among the ranks of Special Operations to bolster the SDV team. It's extraordinarily hard work, requiring an additional eight months of advanced diving and technical instruction. Most of the guys who became SEALs in my era just wanted to jump out of planes and shoot bad guys. The SDV team did not appeal to them—it was more like being an astronaut than being a shooter. Childhood dreams of flying to the moon notwithstanding, I was pretty much the same way. But I had spent a lot of time chasing and shooting bad guys and I had seen a high operational tempo (op tempo) for someone of my age. No generation of commandos since Vietnam had seen the amount of sustained combat action that our generation of troops had seen—and certainly not in the SEAL teams, which traditionally had existed on the outskirts of the U.S. Special Operations community. Before our era of fame and fortune (after Bin Laden), SEALs were considered a niche maritime commando outfit from the '70s through the early 2000s, generally looked down upon by its more prestigious Army counterparts. It's a fascinating history all on its own that I won't delve into here, but you can explore any one of dozens of chronicles that have spawned since 9/11 that expertly tell the history of U.S. Special Operations Forces.

What I discovered when I reported to Hawaii for training was that the SDV team was largely comprised of older guys who had come up during a very different era—a peaceful pre-9/11 era. In the 1990s, SEAL work was very different than it was after the terrorist attacks of 9/11. It was safer, less intense. I don't mean to disparage anyone who has ever served in the SEALs, but that's simply an indisputable fact. Prior to 9/11, it wasn't a real dangerous or even a particularly hard job to be a SEAL. Sure, it was difficult to get through BUDs, the training was still rigorous, and occasionally you went off on a secret mission somewhere. But it was far from the endless loop of combat and deployments that typified Special Operations in the 2000s.

Some guys—a lot of guys, actually (including yours truly)—wanted that newer version of SEAL life. They craved the action, needed it like a drug until they got hurt or killed or simply burned out, which eventually happens to almost everyone. It's a young man's game. For a while, though, the SEALs attracted a fairly high percentage of men who simply wanted to lead a life of adventure, which is not the same as a life of danger. There were triathletes, surfers, mountain climbers—guys who were physically adept but not necessarily seeking combat. Being a SEAL in those days was, for lack of a better term, fun. You'd go on international joint missions, popping in and out of foreign locales on short-term assignments that carried only modest risk, if any risk at all. The pre-9/11 SEAL was basically a globe-trotting, surfing, scuba-diving, extreme-sport junkie. He was a Special Ops bro. This is not to suggest they weren't dedicated patriots and skilled operators, but very few had seen any action.

By the time I left Afghanistan, the SDV team had a lot of these old-school types who had been camped out in Hawaii for fifteen years or more, surfing and training and generally immune from the op tempo most of the other SEALs were subjected to. That created a bit of a cultural divide amongst an already tribal community. While the "other teams" were sometimes on a fifty-fifty combat rotation (spending the same amount of time in combat as at home), the "guys in Hawaii" were notably absent from that. However, in that isolation, there was also a critical component of knowledge retention that would prove to be vital to the SEAL teams years later. As the numbered and colored teams on the mainland were in constant desert combat for twenty years, much of their gear, tactics, and training evolved to create one of the finest and most fierce counter-terrorism communities in the world, replete with the ability to launch daring rescue missions all over the globe at a moment's notice while simultaneously embedding years-long counter-insurgency and intelligence operations in stone-aged villages in ancient parts of the world. What had atrophied, though, was the incredibly physical and technologically challenging skills that were the entire reason for Navy SEALs to exist—complex maritime and underseas tactics.

It may sound crazy, but the truth was that by 2012, most Navy SEAL teams spent less than 5 percent of their time preparing and training for maritime operations simply because they had to prepare for the fight they were in: desert combat. Over the course of a decade, with retirements, casualties, and transitions, our institutional knowledge of fighting on and under the sea had atrophied. The last remaining repository of expertise was a small and highly specialized team in Hawaii that, although considered a sleepy command, carried out some of the nation's most important missions. As I was rotating out of my East Coast assignments, there was a push to bring some new blood into the insular diving command. I was fortunate enough to be a part of this.

That's how I ended up in Hawaii, flying in a minisub during a training mission that did not go quite as planned. Like sharks, submarines never stop moving in water. The ballast on a submarine is never neutral; it is always rising or falling depending on speed, so the slowest a submarine can go is a couple of knots. Any slower, and it will either rise to the surface or fall to the bottom and sit. Therefore the "mating" exercise, where the mother sub launches and recovers the smaller SDV, is sort of like an underwater aircraft carrier routine. You fly the minisub into a big hangar on the back of the moving Virginia-class submarine, and then the door is shut behind you, and water is drained. You do the same in reverse to launch the SDV for a mission. During this training exercise in the Pacific Ocean, the dive team crew had just hooked up our minisub with cables attached to a winch when the mother sub lost its "bubble," meaning its ballast, causing the sub to begin a dive—not a totally unusual procedure, but one that requires immediate action on the part of the minisub pilot and navigator. The minisub, it should be noted, is not a dry sub. It's a six-man wet sub—more like an underwater convertible—so we were out and exposed to the elements, still wearing our breathing gear, and suddenly sinking to the bottom. Very quickly. The mother sub could safely submerge to eight hundred feet, but at that depth, without the protection of a pressurized cabin or dive suit, the crew of the minisub would have been crushed.

We had only one option: pull the emergency breakaway on the cable before the mother sub descended too far. But, of course, once we cut the cable, we immediately shot to the surface, far more quickly than is prudent because a rapid ascent can lead to decompression sickness (also known as "the bends"), a potentially serious or even fatal condition caused by a rapid change in pressure outside the body. This can happen in many settings—high altitude climbing, unpressurized air travel—but is most common in diving. When diving with compressed air tanks, the diver takes in oxygen and nitrogen. The former is used for breathing, but the latter is mostly dissolved in the blood. When a diver returns to the surface, the water pressure gradually decreases, allowing the nitrogen to dissipate in blood and tissue in the form of tiny bubbles. Ascending too quickly, however, can cause a dangerous buildup of gas created by the bubbles (think of what happens when you open a shaken can of soda).

I knew it was happening as we raced to the surface—I could feel the pressure building in my ears and lungs, the heaviness in my joints. It was likely that we were going to suffer from the bends; my hope was that it would not be a particularly serious or debilitating case for anyone. Upon reaching the surface, I had joint pain, but like all my previous wounds and injuries, I kept it to myself in the hopes that I wouldn't be injured in a way that would remove me from duty. I actually didn't say anything about it for more than a day, hoping the bubbles would absorb back into my body and all would be fine. I had been shot and wounded by shrapnel and IEDs and had never reported it before, not wanting to be removed from my team. I figured I could do the same here. But the next day, I was on a bike ride with my four-month-old daughter in tow, and I realized I couldn't keep doing things the way I did as a young single guy. This was my first injury as a father, and knowing that I was an irreplaceable person in her life made me realize I'd better start taking better care of myself.

I called the medical team and reported the incident. Several hours in a decompression chamber reduced the most serious symptoms, and before long I felt basically normal. But I wasn't normal. As it turned out, the decompression sickness had left me with a tiny hole in my heart. It could have been much worse. I felt fine. I was alive, healthy, and seemingly unencumbered by any long-term injury.

But I was, in the eyes of the Navy, damaged goods.

There would be a period of recovery and evaluation, I was told, before I could return to active duty.

"How long?" I asked.

"Hard to say. Could be a year, could be two years. Maybe longer, maybe shorter. Maybe never."

The Navy gave me the option of fulfilling what is known as a "staff tour," in which officers who have been serving as operators on deployment fulfill a professional development tour, typically at the Pentagon, the White House, or the U.S. Naval War College, among other possibilities. Professional development tours usually are done between operational tours—payback for the more exhilarating work of combat deployment—and most of my colleagues had already completed theirs, but I had been on a hamster wheel of deployments for several years. Now, suddenly, I had all the time in the world. Professional development tours are important to the grooming of future high-ranking officers, but I was no longer interested in following that path if I didn't know I could return to operational status. If I couldn't be out in the field, leading from the front, then it was time to consider doing something else. I had put in my time; I was free to go if that was what I wanted.

My wife, Carmen, also a Naval Academy graduate and a Marine, was finishing up a staff development position of her own. In the five years that we had been married, we had never spent more than a few days together in one stretch. We had recently had our first child and were thinking hard about what we might want to do with the next chapter of our lives. We both were somewhat discouraged with the way things were going with the Afghanistan campaign—the lack of resolve, the cycle of drawdown, build up, drawdown again. It was hard to see your friends sacrifice so much, sometimes everything, while the bureaucrats dithered it away.

We wanted to control our own destiny and had kind of grown frustrated with the military. It sounds perverse, but it's true: we both loved deploying, we loved leading people, and we loved the mission. We both loved being in a war zone. But what we hated was the military's constriction of your life and your path; the fact that the trajectory of your career was not determined by your merit, but rather by a giant 1950s-era corporate promotion system. What frustrated me the most about the Navy was that promotions seemed to be determined not by your ability or success in the field, but rather by all sorts of other extraneous shit—all the garrison crap and ass kissing, where you got your master's degree, how long you had been around, and what staff tours you filled. And during the Obama years, it was unquestionably becoming an expectation that as a senior officer you had to toe the line on social initiatives that were being prioritized over combat readiness. I am all about equality, inclusion, maternity leave, and other fundamental rights—both in and out of the military. It's the right thing to do. But I don't care what your orientation or race or gender is—if you can kill the enemy and help us win wars, I want you on the team. The military is not paid to be tolerant; it is paid to win wars. When you take a third of your training time for young troops heading to war and instead of training lifesaving tasks, you indoctrinate people in diversity and inclusion, you are compromising their ability to come home to their families. This sounds myopic and naïve, but the simple fact is, as a young officer preparing to lead young men and women into combat, you are the one ultimately responsible for whether that "kid" comes home to their parents, their wife, or their children. Depriving them of critical time and resources to prepare themselves for combat, in favor of tolerance, inclusion, and diversity training, is incredibly frustrating. But it's not just that the training has to happen, it's the fact that as an officer and leader, you are expected to embody the expectations and demands of the service and evangelize and enforce the rules to the people under your charge. It's not just being asked to go to a different church, put your head down and listen; they are asking you to become a preacher for a different religion. For me, having been wounded, seeing men die in battle, knowing the grim realities of combat, I couldn't embrace that. I was incredibly grateful to have served beside so many amazing warriors and learned from so many great leaders; it was a privilege to be a part of the SEAL community. But my injury also made it clear that it was time to move on.

Carmen and I were both tired of the corporate analysis of the military. You couldn't leapfrog the chain to get promoted based on ability (not that either of us were in a hurry to get promoted out of tactical leadership roles); you couldn't laterally move to something else more creative. It was a tough decision for us, to be sure; we cherished the comradery and mission of the military, but we also tired of being cogs in a big wheel, with little to no ability to make an impact.

So, I guess if it's possible for a life-threatening injury to be considered a blessing, then maybe that's what this was: a forcing of our hand, an opportunity to reexamine our lives. How could we use the skills we had developed? How could we exit the military and still lead a life of meaning and purpose and patriotism—a life of service?

There had to be an answer.

CHAPTER 4

THE ROAD LESS TRAVELED

It seemed as though we had a clear and carefully mapped out plan. A practical, methodical approach to creating a less ordinary life. A "hack," if you will. In reality, we were basically flying blind, guided mostly by youthful exuberance and faith in the idea that, together, we made a pretty good team. But if I could step into a time machine and travel back nearly a decade, I'm pretty sure I'd look at that version of myself and say, "Wow, you're not just arrogant, you're an idiot." That it all seems to have worked out well in the end doesn't change the truth of that observation.

We worked hard, we followed our dreams…and we got lucky.

We also were willing to fail, which I have come to believe is a fundamental part of the process when it comes to building a successful business. Our initial business plan failed utterly; we never made a dime from it. But we did learn from it. We adapted. We moved on. We grew. And that's why I tell everyone who comes to work for us today, "Brace yourself for failure." If you're doing anything worthwhile, if you're challenging yourself, then failure is inevitable. Might be large, might be small. But it's going to happen, and you will learn more from those failures than you will learn from success. We failed so many times with this company…that's why our progress was so fast—we kept failing and pivoting, figuring things out and trying again, never quitting or refusing to open another door. I can encourage my team to keep pushing until failure because that's when we'll figure out where our boundaries are. And that will tell us how to adjust and grow. If we insulate ourselves and only act on what we think will be successful—if we take the easier, well-trodden path—we're just creating an echo chamber of customer feedback, of scientific and business feedback that supports conclusions that we've already made instead of experimenting, pushing ourselves, and testing our limits.

Not that we knew any of this in late 2013, as our time in the military was growing short, and we were trying to figure out what we were going to do with our lives and where we were going to do it. Carmen and I had only decided that we wanted to control our own destiny; we wanted to be in charge of our own lives, even if that meant a more spartan existence. Don't get me wrong—neither of us had any romantic notion about living off the grid or raising a family without adequate financial resources. At the same time, however, we had both decided that we did not want to be at the whim of some gigantic multinational corporation. Nor did we want to work for the government, at least not directly. We had been doing that for the better part of the previous decade.

We weren't wealthy, but we did have resources. Carmen and I had lived quite frugally during our time in the military, spending a lot of time deployed, accumulating savings, taking advantage of base housing and meals, and of course spending almost nothing while on deployment. So, we had amassed a nest egg of close to $300,000. I also had some money that my parents had been putting away for me since I was a kid. All told, we had roughly $400,000 to allocate toward building a business and establishing a new life…somewhere. I was not a rich kid, so I realize that is not exactly chump change, but we knew it wouldn't last long if our plans went awry. And we doubled down on our dreams by prioritizing *place* over almost all other criteria.

From the beginning, Montana seemed like the ideal spot. I was raised in the upper Midwest and had visited Montana. Both Carmen and I had also experienced extended training sessions in Montana. We were avid outdoorspeople and loved to hike, camp, and ski. We knew the region well, loved it, and became somewhat obsessed with the idea of settling there and raising a family. It seemed like a place that was consistent with our values and the way we wanted to live. We were still based in Hawaii (Pearl Harbor) at the time but had accrued enough leave to embark on an extended reconnaissance mission. We flew to Montana and began driving all over the state, surveying various metropolitan regions, not only to see if any of them felt like "home," but also to examine more practical matters like housing, transportation, and the overall business climate of a given city or town.

By this time, the seeds of what Montana has become today—a state with a thriving economy and a serious housing shortage thanks to an influx of coastal money and second home-buyers—had already been planted. But the situation was not nearly as dire as it is now. Carmen and I visited the usual Montana hot spots: Kalispell, Butte, Whitefish, Missoula, and of course, Bozeman. For many reasons, Bozeman struck me as the ideal landing spot. It was beautiful, with snow-capped mountains rising above the city into a seemingly endless blue sky. There were myriad opportunities for outdoor activities practically in your backyard. Land was hardly cheap, but it was still within reach, even if you weren't a tech billionaire with a private jet. Beyond that was the presence of a significant educational center, Montana State University, which had a reputation for producing outstanding students in technical fields like engineering and computer science. I had only the vaguest idea of what my fledgling business venture would look like, but I at least had the foresight to envision a time when that venture might be big enough to require the services of some smart, young, homegrown college grads, and what better pool than Montana State? (Young interns would ultimately be the fuel for our budding enterprise.) Finally, we needed access to an airport with precision approaches, maintenance, hangar space, and reasonable commercial services that would support a growing business. Bozeman Yellowstone International Airport and its neighbor to the south, Ennis Airport, fit the bill.

But what exactly would this business look like, the one that would eventually recruit local college grads and require the services of a sophisticated metropolitan airport? Well, it took some time—and trial and error—to figure that out. When I left the military, the housing collapse of 2007–2008 was not so far in the rearview mirror that it had been entirely forgotten, nor had we completely shed the residue of the recession that predictably followed. The economy was somewhat stable, but it was nothing like the rocket ride in the latter part of the decade. I wanted to be of service in some way, and I figured one of the best ways to be of service was to help create jobs. Then, as now, there was a well-trod path from Special Operations—particularly for a Navy SEAL who had graduated from Annapolis—to the lucrative world of finance. It would have been easy for me to reach into the pipeline and land a job as a hedge fund staffer, investment banker, venture capitalist, or corporate staffer at any number of major corporations. I could have gone straight from the Navy to Wall Street. Plenty of officers with resumes similar to mine did precisely that, and it worked out well for them. They worked hard, success came quickly, and they became millionaires who never had to worry about money again.

I could have walked into almost any job I wanted on Wall Street. Goldman Sachs, Morgan Stanley, J.P. Morgan. It was just out there waiting. All I had to do was pick up the phone. Or answer the phone. Same with Carmen. We were marketable people with high pedigrees. But we talked it over and decided that we did not want to be in the rat race. Number one, we didn't want to live that lifestyle. And number two, we wanted our lives to have meaning beyond our paycheck. Yes, we had grown disillusioned by the management of the war in Afghanistan, and we had decided to move on from the military. But that didn't mean we couldn't still serve our country and our fellow citizens. It didn't quell our desire to serve a higher purpose beyond merely making a ton of money on Wall Street. Additionally, we really wanted to live a life close to the land and teach our kids the origins of their food and how the natural world worked.

Graduate school was also an option. The Wharton School of Business, for example, basically had an unwritten agreement that opened the door for any Navy SEAL officer who applied for admission. Several other institutions also offered this honor to transitioning SOF officers, and it's a great service they provide by doing this.

But neither business school nor Wall Street represented the right path for us. Not because we were afraid of working hard. I can assure you that there aren't many things requiring more work than starting a business from scratch. We just didn't want to work for someone else. There had to be more to life than merely cashing a paycheck—regardless of how big that paycheck might be. The very thought of sitting in a car in rush hour in a big city made me nauseous. And the fact that I had an older brother already working in finance only contributed to my trepidation. Not that he was unhappy; he seemed to enjoy his work. But simply seeing what it involved—the life I would lead—gave me night sweats. Maybe it had something to do with the fact that I had been in combat on so many occasions; I had come close enough to dying that I understood the preciousness of life, and rather than playing it safe, I wanted to live each day as if it might be my last. I wanted to have a purpose. I wanted life to be exciting, fun, worthwhile.

Living passionately, though, is not the same as living recklessly. I had a wife and a baby daughter, plus another child on the way. I was accountable for more than just myself, which meant that while it was all well and good to bypass the safe path, a certain degree of ambition and organization was still advisable. Dewey-eyed optimism went only so far. In the entrepreneurial world, it helps to be both a dreamer and a pragmatist.

So, naturally, I bought a plane.

If that sounds a little counter-intuitive, well, it was and it wasn't. There was an idea, a goal, and a plan to make it work. My inspiration was seeing firsthand the aerial resources that were available when I served in a theater of combat—how valuable they were when properly allocated, and how dangerous when things went sideways. The good generally outweighed the bad, and on those occasions, I found myself thinking, *Boy, if I could just find a way to take that capability, make it more affordable and easier to use, and provide it to a broader spectrum of people who might need it, then I might be onto something.* At the time, it was really only the military—specifically Special Operations—benefiting from that sort of surveillance aerial support. But I could envision other applications in the private sector—or as a private sector entity serving private and municipal clients, including law enforcement and the U.S. Border Patrol. All these communities (along with private ranchers whose herds sometimes strayed from designated zones and needed tracking) could use this type of service in some capacity, so why not start a business specifically designed to offer them surveillance and aerial support?

And that's what we did.

★ ★ ★

In February 2014, I notified the Navy of my intention to resign in September, and Carmen did the same with the Marine Corps. That left us with seven months to find the aircraft that would be the centerpiece (okay, the *only* piece) of our new business venture, as well as figure out where we were going to live and how we were going to pay for everything. In the meantime, I was still in the Navy and Carmen was still in the Marine Corps, even if our enthusiasm had waned to the point of disinterest. It was a long slow exit, as it often is when you separate from the military.

I couldn't be on a SEAL team anymore, so I had become a logistical support officer for the team, reduced to doing the kind of boring staff work that I never wanted to do. In fairness, though, there was nothing else left for me to do. I was still an officer in the U.S. Navy, and I had to do something to earn my pay for the next seven months. My job was to be a safety officer—observing and grading SEALs while they trained in Pearl Harbor and helping to coordinate support operations for active teams doing missions. It wasn't what any SEAL wanted to do, but I was grateful to the SEAL teams and military community for the experiences

I'd had, so I tried to do my best at these somewhat mundane tasks and continue to be of service on my way out.

One night—and I mean, the *middle* of the night—while standing on the beach, helping to grade a diving exercise, I had some time to kill. I was standing in the darkness, watching the moonlight reflecting off the waves, thinking about the future. I was only twenty-eight years old, but I was in such a hurry to get going, to move onto the next phase of my life. A few hundred yards away, some Navy SEALS were exiting a mini sub and preparing to sweep the shoreline. My job was to critique their performance, but my mind kept drifting. It was a long exercise, and it would be hours before we were through.

I pulled my cell phone out of my pocket and began scrolling through the directory of contacts. I stopped at the name *Sam Beck* and smiled. Sam, born in Korea and adopted by American parents, was one of my oldest friends—we had met in middle school, become good friends, and remained close despite time, distance, and the fact that we had very different personalities and political beliefs. But we also had a lot in common. Sam had gone into the Navy following his graduation from Columbia University with a degree in engineering. Smart and ambitious, Sam was an active-duty nuclear submarine officer. His role in the military was quite different from mine, but no less competitive. The funnel that leads from basic training to piloting a nuclear submarine, like the one that leads to becoming a SEAL, is extraordinarily narrow, the competition for slots ferocious (as it should be—you only want the smartest and the best leading submarine crews carrying nuclear warheads). It is, unquestionably, an elite community of officers.

That said, it's not for everyone, neither by aspiration nor qualification, and even the most accomplished officer and engineer can grow weary of sitting in a tube beneath the ocean's surface for months at a time. Despite its importance, it can be boring, repetitive work; most people eventually tire of it and look to move on. Maybe, I figured, Sam had evolved into one of those people.

If you're lucky, you'll have a chance to develop the kind of friendship that I had with Sam, the kind that transcends differences and survives gaps in contact. We were both busy, rarely home, working jobs that did not encourage keeping in close contact with people outside our military environment. There simply wasn't time or opportunity. For me, on deployment in Afghanistan, weeks would pass without phone calls to anyone back home. Given a chance to connect, I reached out to my wife. In that world, you prioritize; very few people penetrate the bubble. I'm sure it was the same for Sam.

When I dialed his number that night, it had been at least a year since we'd spoken, probably two or three years since we'd gotten together in person. But when he picked up, it was like no time had passed at all.

"Hello?" Sam said, in the same deep, measured voice he seemed to have had forever—even in middle school.

"Hey, Sam. It's Tim."

"Yeah, I know. What's up?"

Our calls were so infrequent that when they occurred, there was a tendency for one of us to wonder if something was wrong—if some tragedy had befallen either the person making the call, or a family member or a friend we had in common.

"Nothing, just checking in," I said. "Been a while."

"Yeah, it has. Everything all right?"

"All good. You?"

"Same."

There was a long pause before I jumped into the meat of the conversation.

"Listen, Sam, you don't know about this, but I got injured recently—"

"You okay?" he asked, cutting me off.

"Yeah, yeah, diving accident, but I'm fine now. No worries. I got lucky."

"That's good."

"Uh-huh. Thing is, I'm not gonna be a SEAL anymore. No more deployments, no more missions, and you know how I feel about staff work."

He laughed. "Right. So, what's next?"

"Well, I've decided to leave the military. Carmen and I are both getting out in September, and we're thinking about setting up a little business."

Another pause.

"What kind of business?"

I gave him the elevator pitch, which took all of about two minutes and was decidedly lacking in detail. There would be a plane, I said (although I hadn't purchased one yet). There would be a home and a headquarters in Montana (although we had procured neither of those things to date). There would be a business built on the notion of providing surveillance and aerial support for government entities, the military, and law enforcement (we did not have a single client).

"I think it's a pretty good idea, and it could be an interesting little company," I concluded, sounding like I was trying to convince myself as much as I was trying to convince Sam. "Anyway, I was wondering if you might want to help me get it off the ground. But I have to be honest. It's a total startup. I don't have much money. Actually, I don't have any money, other than what I'm putting into the business, so I can't pay you. I don't know what your plans are, how long you want to stay in the Navy, but I wanted to throw it out there and see if you might be interested because I think you'd be a great partner."

There it was. I had just offered a guy (one of my closest friends, no less!) with a world-class job piloting nuclear submarines the opportunity to chuck his career and work for free in a venture that had no concrete business plan nor any capital beyond what I was putting up from my own savings. How could he possibly pass it up?

"Interesting," Sam finally said, after yet another interminable pause.

Sam was not much of a talker. Every word mattered to him. Every word was considered before entering the atmosphere. "Interesting" to

Sam was not merely a place holder. It meant exactly what it was supposed to mean. He found the pitch…*interesting*.

I let the word hang there for a while before responding.

"Is that a 'yes?'"

"Montana, huh?" Sam said, in what seemed to be a deflection.

"Yeah, Bozeman. Or just outside Bozeman, anyway. I know it seems kind of random, but it has a lot to offer, and it's a great spot to raise a family."

"You know, I have an uncle who lives near Bozeman."

"No shit?"

"Yeah, I've been there a few times. Nice little city."

"And great skiing." Sam wasn't much of an outdoorsman, but I figured I'd throw that in anyway.

"Uh-huh."

Another break in the conversation before Sam asked, "When are you going to start this thing?"

"Carmen and I will probably get there in the fall sometime. I've got a lead on a plane and I'm putting together the financing and paperwork now. It's going to take all the money I've got, but I think we can make it work. Thing is, I need help with engineering, so I need someone who's smart—a lot smarter than me…"

I waited for Sam to laugh. He didn't.

"Anyway, that's the sales pitch, buddy. What do you say?"

"Okay."

"Okay?"

"Yeah. I'm in."

"Just like that?"

"Why? Is there more?"

"Ummm, no. I mean, not right now. I just figured you'd want to think it over."

"Nope," Sam said. "This sounds great. When do you need me?"

"What do you mean?"

"When do I need to be in Bozeman?"

★ MUDSLINGERS ★

Sam was such a careful, methodical person that his eagerness seemed almost out of character. I knew for a fact that my sales pitch hadn't exactly been electric, so some other force had to be at work. *Maybe*, I thought, *Sam is just looking for a change.* Maybe, like me, he'd simply had enough of military bureaucracy. Regardless, I figure I owed him one last chance to hit the eject button.

"You're good to do this?" I said.

"Uh-huh."

"I mean, you did hear me, right? There's no money. Not now, maybe not for a while. I'm just sort of making this up as I go along."

He laughed. Sam didn't laugh often—you had to really earn it—so I took that as a good sign.

"No, I understand. I totally get it. No money, no real plan. Sounds great."

"When do you get out?" I asked, circling back to a question I should have asked earlier.

"I'll give my notice to command tomorrow," he said. "I can probably be in Bozeman around…November? Would that work?

"Yeah, that's perfect."

"Great. I'll see you in the fall."

"Sam?"

"Yeah?"

"Just to be clear…we're really going to do this?"

"Guess so."

★ ★ ★

Here's the thing about commercial and private aviation. It's expensive. Like, extraordinarily expensive. Monopoly money expensive. I thought I knew this before I got into the business, but as with any new venture, the reality of it is never quite what you anticipate. Money goes out the door so fast that it's easy, if not likely, to go bankrupt before the business is technically even off the ground (so to speak). There have been times over the years when it felt as though it would have been more fiscally prudent to simply set fire to great piles of cash in the backyard than to continue investing in aviation. Either way, money simply vanished into the wind. But that is the nature of the business, and you either learn to deal with it or you give up. It almost helps to be a bit naïve and idealistic, to think that you can overcome the odds.

Finding a small plane isn't as challenging as you might imagine. There is no shortage of sellers, and of course the internet has made it convenient for buyers and sellers to connect. I became a vigorous scanner of the marketplace and eventually tracked down what seemed to be a reasonably good investment: a forty-year-old Twin Commander 500 that had been used by the National Oceanic and Atmospheric Administration (NOAA) to conduct atmospheric and ice cap surveys in the arctic. Now, if you know nothing about aviation, you might think that a forty-year-old plane sounds like a risky investment. And it is, to some extent, but not necessarily because of its age. The fact is a plane that has been in service for four decades is far from being a geriatric member of the fleet. If properly maintained and serviced, a forty-year-old plane is simply mature and should still have plenty of good miles left in the system.

Of course, that is a very big "if."

Not that I really had much of a choice. Like most people who buy their first airplane, I experienced a bit of sticker shock when I first began shopping. Even a modest little Twin Commander, forty years of age, maintained to a level that would at least inspire enough confidence to climb into the cockpit, was going to cost approximately $150,000. And sure enough, that was the asking price for the plane that interested me the most. It was located in Indiana and had been targeted for liquidation, meaning it was no longer going to be part of the NOAA fleet. This was not necessarily a bad thing or a sign that anything was wrong with the plane. Government agencies often have the wherewithal to invest in new equipment, even when older equipment is still perfectly serviceable. Moreover, they have both the resources and the incentive (potentially punitive oversight) to take exceptionally good care of their fleet. It was reasonable to believe that a NOAA plane would be in sturdy condition relative to its age.

There were other reasons for choosing this particular Twin Commander. Modifying a plane for a client's particular needs can be a laborious and expensive process. Anyone can take a hacksaw to a plane—or hire someone to handle the hacksaw—but with each modification comes a set of rules and regulations that must be met and adhered to before the plane can be put into use. The FAA carefully monitors and regulates all modification procedures, and the journey from drilling the first hole to flying for the first time can be long, arduous, and costly. It's not unlike modifying a structure for commercial use. There are permits that must be obtained and regulatory bodies that have a responsibility to make sure everything is being done according to code, and the power to demand changes or even assess fines if shortcuts are being taken.

The Twin Commander 500 I had found in Indiana already had a bunch of sensor holes in it from its time in service for the NOAA because they were flying with all kinds of different cameras and other equipment on board. In the world of aviation, a plane with supplemental type certificate (STC) approved equipment is a huge deal in terms of both time and money.

Since STCs and approved engineering are a core part of airtanker development and therefore foundational to aerial firefighting, I will briefly explain what they are. The design of an airplane must be approved by the regulator—the Federal Air Administration (FAA) if you're a manufacturer based in the U.S., Transport Canada for Canada, etc. The Type Certificate Data Sheet (TCDS) is like the approved architectural plans for a building. Once the engineers submit it for approval and the city or county issues a permit, you have to stick to that design down to the brick and outlet. Any deviations have to be approved. It's the same and even more stringent for airplanes. Once the FAA approves the design, any deviations usually require an STC, which is an addendum to the TCDS that allows owners and operators to modify their aircraft for a specific purpose. Even if it's something as simple as adding an extra power outlet or antenna, those procedures could cause an inflight fire, electrical overload, or depressurization, all things that can quickly turn deadly. Therefore, finding a plane for sale that NOAA modified with every conceivable modification would potentially save us a lot of money and time down the road.

A big part of our plan involved the use of surveillance equipment, so I knew that this plane would be useful and practical in almost any application. It was a $150,000 plane, which is a lot of money, to be sure. But it was $150,000 for a flight-ready plane, as opposed to a plane that would end up costing $250,000 after modifications. I did my due diligence—checked out the plane personally, hired my childhood flight instructor to thoroughly vet the aircraft and take it for a test flight. Then came the toughest part—reaching out to my parents and asking for access to the money they had been putting away for me since I was a kid. Ostensibly a college fund, the money had been set aside for use as an adult because there were no costs associated with attending the U.S. Naval Academy. The original stipulation was that I would not be able to access the funds until I turned thirty years of age. I was twenty-eight at the time. So, I called my parents and explained what I was trying to do, asked if we could accelerate the calendar by a couple years, and they graciously agreed. I have no doubt they were deeply skeptical of my harebrained idea, but they and my brother were exceptionally supportive. In addition to the $100,000 loan, they offered me plenty of free advice, which as anyone knows in family business, can go both ways. But nothing would have moved forward without them.

In the beginning, there was no expectation that our little business venture would ever be anymore more than that: little. One plane, with me as the pilot (I had made sure to keep my license up to date), my buddy Sam overseeing IT and engineering, and maybe a mechanic and an office assistant. Over the next few months, I tried to narrow the scope of our business as much as possible. Initially, I was thinking our most likely customers were folks involved in any of a broad spectrum of

★ MUDSLINGERS ★

public safety endeavors—from law enforcement (local or national) to search-and-rescue teams. Police typically use helicopters, which are very expensive to operate and expensive to buy, and because they operate at low altitude, they don't cover a lot of range. Helicopters are good for a lot of things, but not so good for others. I thought we could provide a much lower cost alternative, one that offered more visibility and range and a smaller footprint. Basically, we'd offer surveillance of whatever could be of value to the public safety market. The possibilities seemed limitless.

Let me give you an example. One of the first things we hoped to do was install an infrared camera in the Twin Commander (this little camera idea would eventually become an entirely separate business for us). How is this useful? Let's say you have a missing hiker in the mountains outside of Los Angeles (which is much more rugged than you might imagine, and in fact, hikers get lost there with some regularity). Law enforcement and other agencies will typically dispatch teams of rescue workers to scour the region on foot, which is slow and labor intensive. Support will be provided by helicopters, which are helpful, but they can only stay out for so long (because they use a ton of fuel) and are limited by altitude. A plane can stay out longer and cover more ground, and, if it's outfitted with an infrared camera, it can quickly scan the area for human activity.

Another example: patrolling the border between the U.S. and Mexico or the U.S. and Canada. This could be to provide humanitarian aid, or it could be to provide support for law enforcement. Either way, I figured our services would be useful as the Border Patrol was perpetually understaffed and underfunded. Private contractors are used to supplement the resources of law enforcement and public safety at many levels, and I presumed we'd have no trouble getting work in this realm.

Finally, I hoped to use my exprcrience in the military to secure contracts offering training support across a wide range of applications. I knew from personal experience that there were gaps in knowledge and experience even among the most elite Special Forces units. There just wasn't enough time devoted to training with airborne assets. So we decided that our aircraft could be a stand-in during training exercises, offering the military a low-cost alternative at a high rate of return. I had seen the value of aerial surveillance (in myriad forms) overseas. At its best, utilized properly, it saved lives and facilitated missions. I knew that air surveillance was in its infancy in the modern age and was going to be a focal point of military functionality going forward. The increasing reliance on drone capability—surveillance for tactical strikes that limit the likelihood of casualties—has proven this to be the case. But our business, even in those early days, was built on the belief that aerial surveillance in multiple forms was going to become central to a lot of different functionalities—both in the military and private sector.

But you know what I did not anticipate, or even consider? That our little airplane, and the technology supporting it, would ever be used in the fight against wildfires. Never occurred to me. I had not grown up in the Mountain West and was mostly unaware of the phenomenon of wildfires and their growing impact on both the environment and a huge segment of the population. It wasn't that I didn't care; I was just clueless. And this remained the case even after I first moved to Montana. When you see an aircraft fighting wildfires, you just sort of assume it's an agency aircraft—operated by the U.S. Forest Service, the National Park Service, or a state or regional municipality. I mean, that's the way it works if your house catches fire, right? The local fire department shows up and takes care of things. I figured it was the same with wildfires: a totally closed shop managed by a government agency. I had neither the knowledge nor the inclination to penetrate that world. Instead, I focused on what I knew best. Or *thought* I knew best.

★ ★ ★

In September 2014, after I had officially left the Navy, Carmen and I flew back to Minnesota with our baby girl and checked in with family and friends. Then we packed up a U-Haul trailer and drove to Montana, where sixty acres were waiting for us. Sixty undeveloped acres. The simple and probably sane thing to do would have been to rent an apartment in town while we got the business off the ground. That's what most people would have done, and it was certainly an option we considered and discussed at length several months earlier. But Carmen and I were both deeply familiar with inertia—the way you can get on a particular path and just stay there because it's easy or comfortable. You see a lot of that in the military—people who plan to serve just a few years, but then get a big bonus to re-sign, and before they know it, they've logged at least ten years in the system, and well, that's halfway to twenty and a pretty good retirement, so...

Time gets away. You settle rather than strive. The idea of pushing out of your comfort zone becomes less appealing. We could easily have moved into a pleasant enough place in Bozeman, thrown ourselves into raising a family and growing a business, figuring "someday, when we're ready, we'll buy some land and build a place in the country." That would have been the prudent thing to do. But it's a slippery slope. "Someday" gradually becomes "never," and you wake up at fifty and wonder why you didn't follow your heart. We were determined not to let that happen.

So, many months before moving, we sunk our savings into a down payment on the sixty acres outside of town, got a VA loan to cover the balance, and committed fully to the life we dreamed of. There was just one little problem.

We had no place to live.

A nice plot of land, to be sure, but no structure in which to sleep, eat, or simply stay out of the elements. More choices. Rent a short-term apartment or a room in an extended-stay hotel. Both were minor expenditures compared to what we had already spent on the Twin Commander and the acreage outside of Bozeman. And yet, they felt like an extravagance. Instead, we bought a barn kit and went to work. Carmen and I both had some rudimentary carpentry skills, and neither of us was afraid to get our hands dirty or make mistakes, but we weren't so full of hubris that we thought we could manage the project on our own. We hired a contractor and worked alongside him, building the barn bit by bit as summer gave way to fall and the Montana nights grew cold.

The barn was nothing fancy—once finished, it would have two bedrooms and about eight hundred square feet of living space. Problem was, it wasn't finished when we moved to Montana; it wasn't even started. To minimize expenses and keep our lives small and tidy—living where we worked and working where we lived—we set up a tent next to the stream on our property and called it home for the better part of three months (with occasional nights in a hotel just so we could shower). I suppose that might have been a deal breaker for most women who were taking care of one baby and pregnant with a second, but Carmen was not a typical mother. She was a Marine, tougher than about 99 percent of the people I knew, male or female. She had been on deployment; she had even been in a theater of combat. Camping in Montana? Not really a big deal.

"Kids will be fine," she said. "I'll be fine."

By November, construction on the barn had progressed to the point where we could occupy a portion of the upstairs. We had running water and a working kitchen. The downstairs became a place to store heavy equipment and cars, and it became our makeshift office space.

Along the way, I got a call from Sam Beck. This was probably in late September or early October. I hadn't heard from him in months, but that wasn't unusual with Sam. He was getting out of the Navy in a few weeks, he said, and wanted to make sure my offer was still on the table.

"Absolutely," I said. "When can you be here?"

"Still shooting for November."

"Great!"

"Can you do me a favor?"

"Anything, Sam. What do you need?"

"I gotta find a place today. Can you look at some apartments for me, make sure they aren't shitholes?"

I laughed. "Sure, no problem."

He called again a few weeks later.

"Hey, just letting you know I'm ahead of schedule and I'm going to be in Bozeman tomorrow. Can you help me unload?"

"Of course."

Sam showed up the next day with his entire life packed into a U-Haul. And on that day, Bridger Aerospace was born. I was the chief executive officer; Sam was the chief technology officer. We were a fledgling aerial surveillance business with two partners, one plane…and zero customers.

Around the same time, I was introduced to a pair or fellow veterans who were also seeking an alternative life to the big corporate grind, Tim Cherwin and Steve Taylor. Tim was a former Army Apache Pilot (who I accuse of almost killing me with a Hellfire to this day, even though he claims that he wasn't even in the country). I hired him over a phone call with about as much fanfare as with Sam. He already lived in Montana, but we had never met. Someone told me he was a good guy who had been asking around about a job that would keep him in the state. So, for $1,500 a month, he landed himself a primo position as the chief pilot of Bridger Aerospace! He brought with him the chain-smoking desert rat Steve Taylor, who would become our director of maintenance, also earning wages below the poverty line.

Several things united us all: we were poor, we were war veterans, we all loved Montana….and we had no idea what we were getting into.

CHAPTER 5

THE FIRST FIRE BOMBERS

In the summer of 1953, few residents of Colusa and Glenn Counties in the Mendocino National Forest region of Northern California were aware of the existence of the New Tribes Mission (NTM). They were mostly unaware that each summer, groups of young missionaries would travel from all over the world to the region to take part in a rigorous outdoor training camp run by the evangelical group; or that, for years, their ranks had served as volunteer firefighters in the area, helping whenever and wherever they were needed. But such was the nature of their service, which was, by design, selfless, quiet, and far from risk-averse.

Founded in Chicago in 1942 by Paul W. Fleming, a California native who had served as a missionary in British Malaya, the New Tribes Mission was no stranger to tragedy, although its followers might have considered it merely part of the job. Their ranks were mostly filled by intrepid young men who sought assignments in some of the darkest and most dangerous corners of the globe. Their mission, as they saw it, was to spread the message of God to those who were least likely to hear it, at least in Western form. This invariably took them to places that were difficult to reach and populated by people who did not necessarily take a friendly view toward visitors or outsiders in general.

In November of 1942, New Tribes dispatched its first group of missionaries abroad. Their assignment was to work with natives in Bolivia. Five members of the group were killed the following year. It was determined later that the missionaries had been killed soon after their arrival, the natives striking them down and dismembering them with spears and machetes, fearing them to be dangerous colonizers and outsiders. It took months for the news to reach NTM that their first five missionaries had been murdered in the jungles of Bolivia while trying to spread the word of Christ. What message was God trying to send them? Was this a test of their resolve? Or a sign that they were misguided in their attempts to serve Him? They eventually concluded that God was forcing them to sacrifice in furtherance of their mission and this was the price to pay.

Undeterred, the mission continued its work in far-flung locales, a commitment that yielded occasionally fatal outcomes. In June 1950, the mission's DC-3—the first plane purchased, owned, and operated by NTM—crashed in Venezuela, killing all fifteen of its passengers, which included missionaries and their families. Later that year, in November, a second New Tribes plane went missing between California and Montana on a flight to collect missionaries and take them to South America, a team that included the founder of NTM, Paul Fleming, as well as twenty other missionaries, including seven children. When the aircraft didn't arrive as scheduled, immediate search and rescue efforts were undertaken.

Calls were placed to airports and communities along the route of flight to see if the aircraft had landed elsewhere and was waiting for weather to clear. Or perhaps a mechanical issue kept it on the ground. Hope was briefly restored when a report came in that they may have landed in Nevada. It was quickly debunked. Then a concerning report came from some farmers in Wyoming. A flame had been seen burning on Mount Moran near Grand Teton National Park. In November, this couldn't have been a natural forest fire; the only explanation could have been that it was related to the NTM flight. Perhaps they had crash-landed and were surviving with a fire? Had they struck the mountain and crashed?

★ TIM SHEEHY ★

A search party was quickly assembled, with Paul Petzoldt, the park's climbing guide, as its leader. As a grizzled veteran of the 10th Mountain Division who had climbed the Grand Teton at only sixteen years old, Paul put a small team of volunteers together to scale the mountain and search for survivors. The climb was treacherous, as the team had to scramble over the snow and icy rocks of November. Climbing in the summer or dead of winter is challenging, but ultimately manageable because uniform conditions are expected and you can plan for a summer rock scramble or a winter snowshoe or crampon trek. When climbing in the shoulder seasons, it can be exceptionally difficult and dangerous, as you must be prepared for significant amounts of both rock and snow. Coupled with the unpredictable and capricious weather of November, it was a high-risk rescue by any measure.

All four climbers made it to the reported fire location, climbing through the night at great personal risk, in hopes of finding some survivors. Fanning out to search the reported area, the team struggled through inclement weather and dangerous terrain. Unfortunately, after eventually finding the wreckage in a steep rock pitch, it was apparent that no one could, or did, survive the crash. All passengers and crew had perished in the accident, striking a terrible blow to the NTM family. In just a few short years, they had endured three fatal events, killing thirty-nine members of their church, including children and their leader.

The DC-3 was traveling from Chico, California (a noted aerial firefighting hub) to Billings, Montana (home of Billings Flying Service). The weather in the Wyoming area deteriorated much faster than had been forecasted, as it is apt to do in this rugged and mountainous part of the country. With winds stronger than expected and visibility down to nil, the aircrew lost orientation at some point and failed to maintain altitude to clear terrain. Blustering along at 160 knots, the thin-skinned ole paratrooper plane from World War II struck Mount Moran at 12,900 feet, one of the tallest peaks in the region. It was by a terrible stroke of misfortune that the plane would strike terrain at that altitude.

Just a few hundred feet in either direction, and it would have been clear sailing.

Paul Petzoldt would go on to use this mission as a foundation for the development of the National Outdoor Leadership School, which provides wilderness-based leadership experiences to youth around the world to this day. Given the harsh terrain, the wreckage of the DC-3 remains on Mount Moran, with climber etiquette mandating that all trekkers give proper respects to the dead and not remove or scavenge anything from the site.

Many decades later, New Tribes would become embroiled in controversies stemming from accusations of cultural interference with Indigenous groups, as well as abuse allegations levied at staff members who worked at one of the organization's boarding schools in Senegal in the 1980s and 1990s, which ultimately led to a rebranding under the name Ethos 360. But neither time nor transition can wipe away the memory and undeniable heroism associated with a group of New Tribes missionaries lost in one of the great wildfire tragedies of the twentieth century—an event that led indirectly and improbably to the birth of aerial firefighting.

It was midday on July 9, 1953, when a fire sparked along Alder Springs Road in the town of Elk Creek, just inside the boundary of the Mendocino National Forest. The sudden presence of fire was not surprising given the time of year, temperatures that had consistently hovered in the low to mid-nineties, and hundreds of thousands of acres of dry timber and brush to serve as fuel. The Mendocino National Forest remains one of the most volatile fire regions in the world, with significant elevation changes that foster an unpredictable climate and a varied ecosystem of sap-filled trees, California chaparral, and other plants whose oil-slickened leaves serve as accelerant.

Forest rangers quickly went into action, summoning more than one hundred volunteer firefighters from the region and attacking the fire at its source, trying to prevent a rapid and uncontrollable expansion. Among those enlisted in the fight were twenty-four members of the New Tribes Mission. Within a few hours, firefighters had dug a line and apparently gained a measure of control over the seemingly weakening blaze. By evening, winds had died down and the Rattlesnake Fire, as it came to be known, did not appear to be the type of event that would become notorious in firefighting lore.

But such is the unpredictable nature of wildfires. A spark jumped the firebreak and set off another small blaze in nearby Grindstone Canyon. U.S. Forest Service Officer Robert Powers led a crew of New Tribes volunteers into the canyon to check on the fire. While they were there, the wind suddenly spiked and changed directions, billowing down into the canyon. Within minutes, the original fire jumped lines on three sides of the canyon and began racing downhill toward the unsuspecting firefighters.

As all wildland firefighters know, fire typically burns uphill, but a particular nighttime quirk of Grindstone Canyon is the occasional collision of hot and cold air that creates downdrafts that can cause fire to reverse directions and burn rapidly downhill, creating a counterintuitive and dangerous boomerang effect that will catch an inexperienced crew off guard. Powers and his crew had just finished building a six-foot firebreak around the spot fire in the canyon and were sitting down to eat dinner when someone began yelling from above them, on the canyon wall.

"Run! Run!"

The exhausted firefighters had settled down to relax on two sides of the firebreak—most on the downward side, closest to the bottom of the canyon, the remainder on the mountain side, nearest to the rim. When the men heard the warning cries, it took a moment for reality to sink in. Then they recognized the forest ranger above them and saw flames sweeping across the canyon. They scattered instinctively—some running downhill, deeper into the canyon, others scrambling up the canyon wall. The latter group made it to safety, just moments before the flames engulfed the canyon. The former group, including Officer Powers, ran for their lives, deeper and deeper, trying to outsprint a fire that was rolling downhill at an estimated fifteen miles per hour, jumping from brush to brush, gaining power and speed as it closed the gap to the men and eventually overtook them.

The next morning, rescue teams recovered the bodies of fifteen fallen firefighters, some of whom had apparently died while desperately trying to dig a fire trench with their hands. Others were reportedly found with their arms linked to each other. The men who died in the Rattlesnake Fire left behind a combined ten spouses and twenty-four children. It was, at the time, the greatest loss of life in U.S. Forest Service history and remained the worst tragedy in wildland firefighting until nineteen firefighters from the Granite Mountain Hotshots were killed in the Yarnell Fire near Prescott, Arizona, in 2013.

Compounding the tragedy was the revelation that the fire had not been a natural act, but rather the outgrowth of an act of arson perpetrated by a twenty-six-year-old local man named Stanford Pattan, who claimed in twisted logic that he started the fires in the hope that the resulting conflagration would lead to him being hired as a firefighter. He subsequently was charged with multiple counts of murder but only served a few years in prison before being released. (For readers who might be interested, a comprehensive account of the Rattlesnake Fire can be found in John N. Maclean's book, *River of Fire: The Rattlesnake Fire and the Mission Boys*.)

The Fire Control Officer of the Mendocino National Forest at the time of the Rattlesnake Fire was an erudite Chicagoan named Joe Ely, who held an undergraduate degree from Dartmouth and a masters in botany from the Yale School of Forestry. Ely had settled in nearby Willows, California, some years earlier and would ordinarily have been involved in fighting the Rattlesnake Fire, but he was already working another fire in Southern California when the Rattlesnake Fire broke out. The tragedy nevertheless struck him as a uniquely personal event, one that reportedly affected him deeply. It also intensified his resolve to enhance safety standards and protocols for wildland firefighters, and he helped inspire the Forest Service and other agencies to create a program called Operation Firestop, intended to promote innovative methods for fighting wildfires, including the increased use of aircraft.

In the mid-1950s, Joe Ely and a former stunt pilot named Floyd "Speed" Nolta, two men with little in common outside of living in Willows, improbably came together in the spirited, hopeful aftermath of this program and helped pioneer modern aerial firefighting—specifically through the use of airplanes as air tankers and water bombers.

Nolta, who was born into a logging family in Oregon at the dawn of the twentieth century, brought his skills as a mechanic to the Army when he enlisted in 1917, shortly after the United States entered World War I. Stationed at Rockwell Field near San Diego, Nolta worked on planes whose pilots included Jimmy Doolittle, the famed aviator who would go on to receive the Medal of Honor for planning and leading the historic Doolittle Raid on Japan during World War II. Between service in the two wars, Doolittle gained notoriety as one of the great daredevil and long-distance pilots in aviation. Although Floyd Nolta was not even a pilot when the two men were introduced in San Diego, they became lifelong friends, and it's fair to say that Nolta's interest in flight was piqued by his exposure to Doolittle.

Nolta got his pilot's license after being discharged from the Army and eventually made a home in Northern California, where he worked first as a mechanic and then as an agricultural pilot. Crop dusting of cotton fields had taken hold in Texas and quickly spread throughout the West as a convenient and efficient method of controlling pests. This naturally led farmers and pilots to collaborate on other strategies that might facilitate crop growth, and while the development of these strategies happened simultaneously in multiple places, Floyd Nolta was certainly at the forefront in California. He is credited with being the first pilot in California (and likely the U.S.) to seed rice by air in 1928. While there are conflicting claims to this honor (a pilot named Frank Gallison of Merced, California, reportedly did the same thing in 1927, and some reports claim Nolta's inaugural seeding was actually in 1930), Nolta is generally credited with developing a method for speed-planting rice that became an industry model. Nolta attached a seed hopper to the fuselage of his biplane and rigged a device that allowed him to funnel precise amounts of seed and fertilizer from the hopper into a box on the plane's belly. As the plane swooped low over a properly tilled field, Nolta would open the box, releasing seed and fertilizer that was spread over a fifty-foot-wide swath by the wash from the plane's propeller.

While primitive, Nolta's device was nonetheless effective. In fact, in a 2022 article for HistoryNet, California writer and historian Ted Atlas noted that Nolta's system greatly improved rice propagation and that his invention is still used by agriculture pilots around the world, albeit in a modernized fashion.

Along with his brothers, Vance and Dale, Floyd Nolta started the Willows Flying Service in the late 1920s, providing seeding, crop dusting, and other agricultural services throughout Northern California's Sacramento Valley. This included the discouragement and sometimes eradication of pests that would destroy entire fields of freshly planted seeds. The Willows pilots would buzz the fields from less than fifty feet to scare off hungry ducks. There are reports that they were even hired to hunt eagles (normally illegal but allowed with a permit in certain cases) that were preying on smaller livestock in the region. And, yes, the hunting was primitive and dangerous, with one man piloting the aircraft through challenging terrain and his passenger wielding a shotgun or rifle. Less dramatic, but no less vital, were jobs in which the Willows Flying Service was hired by the Forest Service to fly personnel and supplies to backcountry locales or to spot wildfires. Sometimes they even used Nolta's invention to spread seeds for forest regeneration after fire season.

While Nolta was busy acquiring contracts for the Willows Flying Service, Joe Ely was building a quiet, distinguished career with the Forest Service, a path on which he remained even as the U.S. was drawn into war again in 1941. Forest rangers were exempt from military service since their job of protecting forests was deemed critical, especially given the threat of Japanese balloon attacks against the U.S. Pacific mainland.

Nolta, meanwhile, found additional work as a Hollywood stunt pilot before rejoining the Army Air Corps, where he was assigned to the First Motion Picture Unit. His duties included flying planes that would appear in training films and other features designed to boost the morale of both troops and the American public. When Hollywood turned to the story of Nolta's friend, Jimmy Doolittle, for a movie called *Thirty Seconds Over Tokyo* that was based on the Doolittle Raid, Nolta was hired as a stunt pilot. In one sequence, he flew a twin-engine Mitchell B-25 bomber under the San Francisco Bay Bridge.

After the war, Nolta and his brothers reorganized the Willows Flying Service. Among his neighbors was Joe Ely, who had settled in Willows with his family after being named Fire Control Officer of Mendocino National Forest. Nearly a decade later, in the wake of the Rattlesnake Fire tragedy and a fire season that burned more than 140,000 acres in California, Ely became one of the people most deeply affected by Operation Firestop, billed as a multi-agency brainstorming session designed to examine wildland firefighting techniques. More specifically, Operation Firestop, as described in a 1959 article published in *Empire Forestry Review*, was "a one-year study designed to explore certain aspects of mass-fire behaviour, carried out in California in 1954, as a preliminary to the formation of a National fire defence plan for wartime conditions in the U.S.A. The movement of fire in rugged topography, fire calibration, the use of chemicals in fire fighting, and techniques of application were…investigated. It was concluded that chemical fire retardants, such as sodium calcium borate can be effectively used as fire suppressants, in the construction of fire lines and in controlled burning; and that helicopters provide the answer to many problems of hose-laying and equipment transport in difficult terrain."

While it posed as many questions as it answered (which is not a bad thing at all), Operation Firestop can be viewed through the prism of history as a rather impressive example of multi-agency cooperation that led to many of the air operations techniques in fire control, including the dropping of water and firefighting chemicals from the air. Aircraft had been used for some time in wildland firefighting, primarily as spotters starting around 1919 and as transport for the first smokejumpers beginning in the early 1940s. But the use of aircraft as true fire bombers would prove to be a game changer.

One of the earliest and most impactful adoptions of aircraft in support of wildland firefighting was, of course, the use of smokejumpers—highly trained backwoodsmen—to quickly attack new starts deep in the inaccessible wilderness of the Mountain West. Although the exploits of smokejumpers are chronicled thoroughly in other books on the topic, we would be remiss not to briefly discuss the importance of their contribution to the history of aerial firefighting, most notably the Mann Gulch fire in 1949, which was one of the deadliest events in U.S. firefighting history.

On August 5, a new start was reported in Mann Gulch in the Gates of the Mountains Wilderness, in Helena National Park. A fifteen-man crew led by foreman Wag Dodge dispatched in the "Miss Montana" aircraft. After circling the drop zone, which was in rugged high desert terrain, the crew made its jump into a relatively gentle slope away from the fire.

The jumpers met up with a local ranger on the ground who was assigned to the forest and quickly devised a plan for initial attack. They were unable to receive weather updates due to their radio being damaged in the drop (its parachute didn't open so it was destroyed on landing). As the team split up and formed attack lines, Dodge moved to high ground to get a better look at the fire. What he saw concerned him: the fire was cornered in a heavily vegetated drainage area and had begun to "blow up." With the smoke plume intensifying, Dodge ran to his team to inform them that instead of building a line, they were going to take an escape route to safety on the ridgeline above, and then attack from a different flank.

The decision came too late, as it quickly became apparent that the fire was bearing down on the crew. Dodge gave the order to drop all gear and move as fast as possible to the ridge, but by now the fire was roughly one hundred meters behind the crew and closing fast. Dodge made the instinctive decision to stop where he was and light what would become known as an "escape fire," a small fire to protect himself from the approaching inferno. He invited other firefighters to join him, but they ignored him and continued scampering up the hill. Thirteen men were burned alive on that ridge; Dodge survived. The entire chronicle of the Mann Gulch Fire is expertly crafted in Norman Maclean's *Young Men and Fire*. The incident became a seminal event in U.S. Forest Service history and inspired many new tactics and procedures to better protect wildland firefighters—chief among them, the use of escape fires and better radio technology.

Like all tragedies—military, civilian, fire, or otherwise—difficult lessons are learned the hard way and drive change. The Mann Gulch fire represented a quintessential failure in communication and command control. With the team's radio damaged during the drop, the firefighters were devoid of the ability to communicate with outside units for weather and incident updates, or to call for assistance. Their inability to communicate or understand the entire tactical picture ultimately led to their tragic demise.

★ ★ ★

In the spring of 1953, just a few months before the onset of Operation Firestop (and before the pivotal Rattlesnake Fire), an incident that could be described as an "accidental" water drop occurred at Palm Springs Airport in the Southern California Desert, where McDonnell Douglas was testing a prototype of its new DC-7 passenger jet. To mimic the weight of a full passenger load, the plane carried a huge water tank in its belly. Near the conclusion of its test flight, the pilot swooped low over the runway and released the contents of the tank, producing a long, wide wet patch that immediately caught the attention of firefighting personnel in the region. A short time later, with the cooperation of the LA County Fire Department and the California Division of Forestry, a true test was established. Small brush fires were set in a remote area, and a DC-7 loaded with water was assigned a drop. The first pass was only mildly effective—the water hit its target, but not in quantities sufficient to completely extinguish the flames. But once larger valves were installed on the plane, forcing water through at a faster rate, the process was significantly more effective. It was far from a foolproof system, but there was no denying the obvious: an aircraft loaded with water had the potential to be of enormous value in wildland firefighting.

This thesis was tested repeatedly during Operation Firestop, headquartered at Camp Pendleton near San Diego, utilizing an Eastern Aircraft TBM-1C (Avenger) from the vast Hollywood fleet owned by famed stunt pilot Paul Mantz. Much of the flying was done by Mantz himself, and the tests were performed using a tank he built into the plane's torpedo bay and, when that leaked, a giant water balloon provided by the National Weather Service. Mantz and his team made countless drops under a variety of conditions, all of which were monitored and recorded by researchers from Cal Forestry and the University of California Berkeley's School of Forestry.

It should be noted here that the notion of dropping water from airplanes onto fires was not born with Operation Firestop. Various methods had been considered, attempted, and abandoned over the years, for reasons related to safety, impracticality, and ineffectiveness. Dropping a repurposed beer keg filled with water might have seemed like a neat idea at the time, but it was a dreadfully overmatched weapon in even the smallest of wildfires. There is even a documented example of a decorated Canadian pilot and engineer named Carl Crossley landing a seaplane on a lake and using a pipe to fill a tank that he attached to the plane so he could make several repeated dops on a wildfire fire near Ontario in 1945. This, in effect, was the first scooper—more than a decade before the Consolidated PBY-5A amphibians would make their debut in the firefighting world. Crosslcy is credited with developing and patenting a system of fighting forest fires using floats modified to pick up water for the express purpose of dropping it onto fires. But he may have been ahead of his time, as the system was successfully tested using a Norseman-CF-OBJ but not immediately pursued by the Ontario Provincial Air Service (OPAS), which employed Crossley.

The Ontario Ministry of Natural Resources and Forestry (OMNRF) and its firefighting branch, the OPAS, remains one of the world's leading aerial firefighting organizations to this day. In 1923, OMNRF built an airbase along the river at Sault Ste. Marie, literally across the river from Michigan, where float-based Curtiss Golden Fliers were utilized for aerial fire spotting across the province. This would sow the seeds of what would eventually become the CL family of scoopers that were mainly designed and built in Ontario.

"A lot of the early innovations just sort of happened," said Al Hymers, a one-time bush pilot who flew scoopers in Canada and the U.S. for nearly forty years. "It was like pilots sitting around the coffee shop trying to figure out stuff, throwing ideas at each other. That's really what it was. But Paul Cook, who was a bomber pilot in World War II, deserves a lot of the credit. He was director of the Ontario Provincial Air Service in the forties and fifties, and he did a lot to push things forward."

The OMNRF has been aggressively innovating in the aerial firefighting space since the early days. In the U.S., things moved a bit more slowly but began to pick up speed following Operation Firestop. In the spring of 1955, just as the fire season was getting underway, Joe Ely began approaching various pilots and forest specialists about the possibility of incorporating more agricultural aircraft into the firefighting landscape. He eventually received formal permission to pursue a program in this regard from his supervisor, Bob Dasmann of the U.S. Forest Service. The foundation of the program was the use of water drops to mitigate and suppress active wildland fires. In short, Ely hoped to assemble California's first fleet of air tankers. After first taking the proposition to the manager of the Willows Airport and receiving only lukewarm support, Ely approached the Willows Flying Service, where he found an eager participant in Floyd Nolta.

"All I had to do was remark that he sure had a lot of experience dropping materials out of airplanes onto farms, and did he think he could do the same thing on a forest fire," Ely would later recall. "He said to come back in a week."

They didn't call Floyd Nolta "Speed" for nothing, nor simply because of his prowess in the air. Not one for relaxing or procrastinating, Nolta immediately went to work on modifying a plane that could do exactly what Ely had requested. He cut a hole in the bottom of his 1939 Boeing-Stearman Model 75 Kaydet biplane and inserted a 170-gallon tank into the plane's belly. Then he covered the tank with a hinged flap that could be opened on demand with the use of a cable managed from the cockpit. Thus was engineered the first air tanker intended specifically to fight wildfires.

But what worked in theory would not necessarily work in practice, so Nolta put it to the test on July 23, 1955, at the Willows Flying Service's private airstrip. While Nolta started and managed a small controlled burn near the edge of the airstrip, his brother, Vance, climbed into the cockpit of the modified Stearman, took off, circled back, and released the payload. The fire was instantly extinguished. The test was an unqualified success, and in that moment, as Ely would later write, "The air tanker was born."

One could argue that a more accurate date for that proclamation was August 13, 1955, when Vance Nolta was summoned by the Fire Control Office of the Mendocino National Forest to work the newly sparked Mendenhall Fire on the west side of Bald Mountain. Just as he had during his test flight the previous month, Nolta successfully emptied the tank of his Stearman while flying low over the flames of the newly sparked fire. This, however, was not a test—this was the real deal, and the Noltas' invention performed admirably. Over the course of the day, Vance Nolta dropped six loads of water to help extinguish the fire. After each drop, he returned to the Gravelly Valley Airstrip in Upper Lake, California, where his tank was refilled by the Ukiah Pine Fire Authority. Vance Nolta's efforts were smoothly incorporated into the entire firefighting mission that day. In what would become a hallmark of integrated wildland firefighting protocol, Nolta's drops were conducted safely and in conjunction with ground crews nearest the fire. In other words, he was not a daredevil trying to snuff out a fire on his own, but rather one cog in the firefighting machine.

Admittedly, though, it was a new and dramatically effective cog—one that could change the way the machinery worked—and as such, it drew plaudits not just from those involved in the Mendenhall Fire, but from the broader wildland firefighting community throughout California. Shortly thereafter, Vance Nolta's Stearman, tail number N75081, was officially registered as the first air tanker in history to fight wildfires in the United States. And August 13, 1955, is widely regarded as the birthday of the American air tanker program.

In 1956, under the leadership of Joe Ely, the Mendocino Air Tanker Squad (MATS) became the first officially sanctioned aerial tanker unit in the world. The MATS started with a fleet of eight aircraft and nine skilled agricultural pilots personally recruited by Ely: Ray Varney, Frank Prentice, Lee Sherwood, L.H. McCurley, Warren Bullock, Harold Hendrickson, and of course, the Nolta brothers, Floyd, Vance, and Dale. The Noltas shared the use of their two Stearman Kaydet biplanes, while Sherwood flew in a Tri-Pacer monoplane (with a Forest Service observer aboard). The other aircraft were all U.S. Navy biplane trainers. Although part of one unique crew united in purpose, the pilots flew out of several different airports or air strips in the region.

In those days, "fire season" was a much more reliable term than it is today, with wildfires breaking out in a somewhat predictable pattern across the country. In California, fire season typically began in July and stretched into mid-autumn. This was perfect for agricultural pilots since the ag flying season typically ended in June. What had long been a dormant and sometimes financially stressful time for the pilots now became their busiest period of the year and an opportunity to earn significant money while doing work that was both hazardous and meaningful.

Agricultural pilots were a natural choice to become pioneers of aerial firefighting, as there were obvious similarities between the two types of work. Whether dropping seeds or dusting crops, ag pilots would fly low and slow to get the job done; the same was true of the new tanker pilots, who likewise needed to fly practically on top of their target in order to execute an effective drop. Both jobs, then, were suited to pilots who would not be unnerved by inherently difficult flight patterns in unplanned and improvised operational scenarios.

But to suggest that ag flying and aerial firefighting presented comparable risks would be inaccurate, as the pioneers quickly discovered. If you were looking for a type of flying with a similar degree of risk, you'd have been more likely to find it in a military combat scenario. Agricultural missions were generally flown in fertile valleys under gentle or forgiving meteorological conditions. A pilot would swoop in over level ground, usually under clear skies, spray the crops or drop some seeds, and then circle back to the airstrip to gather another load. Neither topography nor climate played much of a role in the job—if the weather was unfavorable, you simply waited until things cleared up. If Tuesday was bad, Wednesday would likely be better, and the fields were always flat and inviting.

Compare that to the conditions faced by the MATS pilots in 1956 (and still faced by aerial firefighters today, although they are equipped with far superior technology). Aerial firefighting is conducted over varying terrain, but most often, aerial firefighters find themselves in the mountains or rolling hillsides where climate conditions can be frighteningly unpredictable. High winds flow down the leeward side of a mountain and then rise quickly as the air heats up—which can happen in a heartbeat during a wildfire. Tanker pilots have no choice but to fly so low that they are almost scraping the treetops as they approach a fire and their designated target. It's not unusual for a plane to be literally shoved toward the ground as it meets the hot air rising over a ridge. Throw in sometimes blinding smoke and bone-jarring turbulence and you have one of the most challenging scenarios in all of aviation. Under these conditions, the early pilots, in their comparatively lightweight, overmatched aircraft, would try to drop water on precisely the right spot at just the right moment, knowing that as soon as the load was released, their plane would instantly be drawn upward—hopefully in time to clear the ridge ahead.

To say that this took immeasurable skill would be an understatement. It also required timing, experience, courage…and a fair amount of luck. The slightest mistake, mechanical malfunction, or meteorological surprise could be enough to prevent the pilot from getting back home. But that was the nature of the job. It wasn't for everyone, and it certainly wasn't for the faint of heart. Whether you were part of a ground crew or flying an air tanker, there was risk involved in fighting fires. But it was a necessary fight—one that became somewhat less one-sided with the advent of the first air tanker crews. At least the MATS pilots were paid well for their work—approximately $60 per flight hour. If that doesn't sound like much, well, this was 1955 dollars, which, adjusted for inflation, translates to about $700 an hour today. There is no way to put a price on a human life, of course, but the MATS pilots eagerly accepted the risks of the job. It may have been more challenging than agricultural flying, but it also paid better. Both factors were part of the appeal.

As testament to Ely's foresight, the MATS pilots were astonishingly busy in that first fire season, getting summoned to work a dozen fires in the month of August alone—their first month in existence. Before long, they were getting calls to work fires all over the state of California. A command post was established in Willows, from which air tankers could be dispatched when someone called their emergency number. The early agricultural planes used by the MATS crew were not outfitted with communication equipment until the 1957 fire season, so the pilots would receive both their marching orders and directions while filling their tanks at the airstrip.

The lack of communication obviously made the aerial firefighting effort more challenging. Today's fire zones are a carefully choreographed dance involving a dozen or more aircraft and scores of ground firefighters all in constant communication. Before a tanker drops its load, it is guided through the drop zone by an air attack that serves as an air traffic controller in the sky. There is some question as to exactly when the first observation planes were utilized, but there are early examples of a radio-equipped Cessna serving as a drop coordinator in the field, guiding the air tankers to their appointed spots and ensuring that traffic flowed smoothly. But prior to 1957, the tanker pilots were, in a very real sense, on their own and flying on instinct.

The MATS pilots served as a test case for the viability of air tankers in wildfire suppression and control, and by all accounts, they passed impressively. In one Forest Service study published at the end of the 1956 fire season, it was reported that MATS was involved in twenty-three Forest Service fire events. Aircraft, specifically tankers, were determined to be a "deciding" factor in controlling or extinguishing fourteen of those fires. In four others, the tankers were considered a "significant" factor, and in four they were deemed "insignificant." In one fire, the tankers were determined to be a "hindrance" because they inadvertently extinguished a backfire that ground crews were using to prevent further spread of a fire. Overall, those numbers were sufficient to embolden not just the MATS, but advocates of aerial firefighting throughout the West.

That inaugural season also saw the first examples of retardant being used in fighting wildfires. While tankers filled with water represented a massive step forward in firefighting, they were not flawless. Depending on weather conditions, accuracy, and the intensity of a fire, repeated dumps of a couple hundred gallons of water from a biplane might snuff a fire completely…or be utterly useless. More likely, the result would fall somewhere on the broad spectrum between those two results. But this much was certain: the bigger and hotter the fire, the less effective the use of pure water was. In some cases, ground crews observed water evaporating before it even reached the ground! (This is less likely to happen in modern aerial firefighting, since we have scoopers attacking fires earlier and carrying much larger loads.)

In response, pilots sometimes filled their tanks with a mixture of water and sodium calcium borate. The chemicals not only slowed the evaporation of the water as it fell from the sky, but also proved effective as a ground slurry, since its melting point of nearly two thousand degrees was roughly twice as high as the typical ignition point of a wildfire. This was the beginning of chemical retardant use in aerial firefighting. Whereas water was effective almost exclusively as a suppressant—that is, when it was dumped directly on a fire—chemical retardant could be used to help *manage* a fire. Rather than releasing their payloads only on the flames, the tankers would drop retardant around the perimeter of an active fire. The chemical mixture was viscous and sticky, so it would cling to trees and brush and grass, creating a barrier that made it much harder for the fire to spread. At the same time, the tankers could continue to dump water directly on the flames. In conjunction with ground fighters, this created a multipronged attack that became the model for modern-day wildland firefighting.

The use of retardant led to the early tankers being dubbed "Borate Bombers." The nickname persisted for decades despite the fact that borate was actually phased out within a couple years, owing primarily to the fact that once on the ground, the chemical had a tendency to blend in with the vegetation it covered, making it difficult for pilots to know where they had already dropped retardant, resulting in significant repeat coverage and waste. Sodium borate was replaced first by Bentonite, a mineral derived from clay, and later by Phos-Chek and Fire-Trol, retardants that consisted of a blend of phosphates, sulphates, and other chemicals with both cooling and extinguishing properties. A variety of chemical retardants soon flooded the market, and some are still in use today. They all basically work on the principle of cooling and slowing or redirecting a fire through chemical means, and all include some type of dye or other coloring so that the retardant is easily visible both during and after a drop.

Once established as an effective tool, aircraft quickly became integrated into the broader picture of wildfire management. The MATS program in Willows expanded, and some of the squad's pilots later signed contracts with the California Division of Forestry. The MATS biplanes soon gave way to larger aircraft, such as the TBM Avenger, the Boeing B-17, and the Consolidated PBY Catalina.

★ MUDSLINGERS ★

The PBY, a seaplane waterbomber, was the first amphibious aircraft to be widely used in fighting forest fires. It was the first aircraft to make good on the promise first demonstrated by Carl Crossley in 1945, and thus became recognized as the first "scooper" to be commonly used in aerial firefighting. The PBY only needed a nearby natural water source in order to reload multiple times during a single day, and its increased functionality represented yet another significant leap forward in wildfire management. It remained an industry leader for many years, and even had its moment in the spotlight in 1989, when it was flown by John Goodman's character in the dramatic opening scene of the Steven Spielberg film *Always*, one of Hollywood's few attempts to capture the excitement and danger of aerial firefighting.

As firefighting aircraft—tankers and scoopers alike—grew bigger and theoretically better suited to the job, flying the aircraft became more dangerous. Larger aircraft could carry more retardant or water and their size offered the pretense of protection, but they could be unwieldly if not downright unmanageable when trying to navigate through unpredictable conditions. This presented a conundrum that engineers and pilots still wrestle with today: lighter planes carry less payload, but they are nimbler when dancing through canyons, blinded by smoke and buffeted by turbulence. The simple truth is that the best plane is one that can get the job done and still deliver its pilot safely back to the fire base.

★ ★ ★

The first recorded instance of a pilot being killed while fighting a wildfire occurred in the summer of 1958. His name was Joseph Anthony, and he was working a fire in Northern California's Sequoia National Park for a company run by Paul Mantz. Anthony was flying one of the same TBM Avengers that Mantz had flown while conducting tests at Camp Pendleton. By the early 1970s, there would be eleven additional fatal crashes involving the Avenger and many others involving a variety of tankers and seaplanes repurposed at the end of their military usefulness. In fact, the majority of aircraft used to fight wildland fires in the U.S. during the early decades of aerial firefighting were ex-military planes of World War II vintage (or older). They had been purchased by private contractors, then gutted to accommodate tanks and other equipment required for use as water bombers and retardant tankers. Although these aircraft were required to pass inspection and meet federal safety standards, their viability was called into question after a number of accidents over the years. Although retired military planes are still used in aerial firefighting, those of older vintage (specifically WWII) have been retired since the early 2000s. This makes sense—by the time they got to their first fire, the old tankers of the fifties, sixties, and seventies had already experienced a lifetime of abuse.

Fortunately, aerial firefighting has come a long way since those early days. While it remains a challenging, dangerous job, it is far safer than it was in the 1950s. There are thousands of aircraft working wildland fires all over the world. Air tankers, once little more than a fantasy, are now as important to wildfire suppression and management as the ground crews that have always manned the front lines. The sheer range and scope of tankers available is so impressive that it's hard to imagine a time when there was literally no help from the sky. From the single-engine air tankers (SEATS) that comprise the smallest subset of tankers, to the Super Scoopers carrying upwards of two thousand gallons of water or retardant, to the Type 1 Firebombers with payloads in excess of three thousand gallons, there is seemingly a tanker for every job. There are even mega-tankers—converted DC-10s and Boeing 747s—that can drop twenty thousand gallons at a time.

Granted, these resources are not always employed for maximum effectiveness, but that is primarily a function of bureaucratic squabbling, budget constrictions, or interagency turf wars. By any reasonable metric, aerial firefighting has traveled light-years since its inception, and much of the credit should go to men like Floyd Nolta and Joe Ely who saw the future in the sky.

CHAPTER 6

CHANGING LANES

"Runway" is a term in the world of business startups. I'd rarely heard the word used this way before we founded Bridger Aerospace and certainly didn't consider the irony of its meaning when applied to an aviation company.

In aviation, of course, a runway is a long and smooth stretch of land, typically paved or blanketed in concrete, from which an aircraft takes off. The longer the runway, the more time you have to get up to speed and rise into the air for a successful flight. The runway is also where the aircraft lands at the end of the flight. In a new business venture, "runway" is used metaphorically to refer to the amount of time, resources, and effort required to reach a point of profitability. It's somewhat counter-intuitive. In aviation, a long runway is a good thing. In business, it's generally a bad thing unless you know exactly what you are getting into and have planned accordingly.

Bridger Aerospace was formed in the fall of 2014 with one little Twin Commander 500 that only needed a very short runway. But the business itself, as I was soon to discover, had a very long runway indeed. And I had not planned accordingly. Finding work for the Border Patrol—our target market in the beginning—and law enforcement agencies proved to be much more difficult than I had anticipated. I had some significant contacts in that world. I had a level of expertise in surveillance. I had a product and a pretty good sales pitch (or so I thought). It didn't seem to matter.

I made two trips to the southern border to have meetings with representatives from a variety of agencies through my old military network: the FBI, Border Patrol, Texas Department of Public Safety, and others. All to no avail. On one of those trips, I was actually stopped by the Border Patrol. And by "stopped," I mean pulled over and ordered to exit the car and stand on the side of the road while my vehicle was searched. I asked why I had been detained and was given no explanation whatsoever. It was an interesting and enlightening experience, especially given the reason for my trip.

Finally, after those trips and months of phone calls, I got an honest response from someone. He was a former SEAL who worked for the Border Patrol.

"Listen, Tim," he said. "I think you've got a great concept here, but you're not going to get anywhere with it. This has nothing to do with you personally, or your background, or even the service your company is offering. It's just that there's not a strong desire to secure the border right now."

This was a problem, obviously. And one I hadn't factored into our business plan.

"You're going to be hitting your head against the wall with this effort. If you want to continue to do it, go ahead. That's up to you. Just realize it's going to be a while before you see any success."

I thanked him for his transparency, but I found myself fixated on the loss of what I expected to be a major client for our services. What a significant miscalculation that had been! But there wasn't time for self-pity or self-flagellation. Instead, we shifted our efforts to the military and law enforcement market with a particular emphasis on providing surveillance support during training exercises.

I was not a salesman, but after a while, I got pretty good at making a sales pitch. And then I got pretty good at figuring out that closing a deal is a lot harder than setting up a meeting to discuss a potential deal. A lot of people were willing to sit down with me, probably because of

MUDSLINGERS

my background as a SEAL officer, but almost no one was interested in hiring us.

The only work we could find was from local ranchers whose cattle periodically broke through fences and strayed into the wilderness. When that happened, the ranch owner had two choices: send out a few employees on foot or an all-terrain vehicle, which was time consuming and ineffective, or hire someone to provide an eye in the sky. That was Bridger Aerospace's primary mission in the early days, and it was far from lucrative. The going rate was about $200 a day, which was barely enough to cover fuel costs. But it gave us an opportunity to test and refine our technology in the wild, so to speak—to implement the sensors and infrared cameras that Sam was developing in real-world applications. It also presented an opportunity to get the word out about Bridger, although not many people seemed to hear it.

By February, our team had expanded to four people. Tim Cherwin and I took turns flying while Sam handled tech development. We did not have a full-time maintenance staff, but instead farmed out whatever work Steve Taylor could not handle himself. Commitment to frugality notwithstanding, we had serious money issues. Cash flow was non-existent, and we weren't even halfway down the runway. At this rate, we'd be out of business by summer. The dedication of my wife, my brother, Tim, Steve, Mike, and Sam were all that got us through those hard few months. In the years since—as we have grown, shifted, pivoted, and expanded—I often look back at that period and remind myself, "If we can get through that…we can get through anything." (This is a common mindset among military folk that is a product of hard military training, which is precisely designed to create this mindset—suffer now to succeed later.)

If that sounds like a stressful time, well, it was. But not to the extent that you might imagine. Maybe it had something to do with being young and idealistic; maybe it was because I had experienced more stressful situations—pertaining to life and death—while on deployment. I can't really explain it, but for some reason I just felt like everything would work out somehow. Sam, Tim, Steve, Carmen, and I were happy warriors and sure that we would find some way to make our vision a reality.

One day at Ennis Big Sky Airport in Montana, a U.S. Forest Service officer named John Agner noticed our aircraft and approached admiringly.

"Hey, I heard from Tim Cherwin you got a Twin Commander. That's our favorite plane for aerial attack."

"Really?" I said.

"Yeah, that's a great aircraft." He paused. "You ever think about flying wildfires?"

The truth was, no, I had not thought about it. Not for a second.

"I thought that was all government work. I mean, I just assumed…"

"Well, administratively, yeah, it is," John said. "But most of the aircraft are under private contract. We're going to be short on planes this summer, so if you're interested, we could really use a Twin Commander to help with air attack." He stopped, pointed at one of the infrared cameras. "I like your tech too."

"Thanks," I said. "We're putting a lot of effort into it. What's involved in…air attack? Is that what you called it?"

He nodded. "Most of the time you'll be used for flying in the mountains, spotting fires."

Spotting fires…. It sounded simple enough.

"Okay," I said. "How do we get started?"

"Give me your contact info and I'll send you the information."

It was only later that I got the full story from Tim about how this "chance" meeting occurred. As I would find out, this budding relationship was the result of a few too many drinks, far too many words, and not enough common sense (this happens in the military as well, it should be noted). A few days prior to our meeting, John and Tim had gotten into a heated "discussion" at the Gravel Bar in downtown Ennis about a topic no one can seem to remember. After some insults were exchanged, some collars ruffled, and some pride damaged (not sure on which side), Tim somehow managed to give him the sales pitch on our business. The rest is history.

Much homework ensued. I became obsessed with aerial firefighting, a subculture whose existence I had barely even recognized. Frankly this was a bit of an embarrassment, given my background in the Navy and my youthful fascination with flight and aerospace. And I was an outdoorsman; I lived in Montana now! How could I have been so profoundly unaware of the importance of flight in the ongoing battle against wildfires? I read stories of the early pilots, the tanker scandals of the 1980s and 1990s, and everything in between. It was like an entire segment of the aviation world suddenly opening up to me, and I wanted to be a part of it. Not merely for business purposes, but because it struck something in me that I hadn't felt in a long time. Here it was—a chance to do important work. Thrilling work. A chance to be part of something bigger than myself. Much bigger. Admittedly, the excitement was tempered by the realization that the clock was ticking and that we had to come up with a path to sustainability. But here, it seemed, was the answer to our prayers, to our ambitions. A way to make Bridger successful while fulfilling our mission of service.

I knew nothing about government aviation contracts, but quickly became immersed in the details. When the contracts were emailed to us, we all got excited. There was an opportunity to make a couple thousand dollars a day doing work that sounded legitimately exciting and useful. I know it probably sounds corny, but the mission was attractive. We wanted to give back. We wanted to be of service somehow. I can honestly say that the goal was not to become multimillionaires; the goal was to create a viable business in a region of the country where we wanted to live and to provide jobs that would support the people of that region while doing work that mattered in some way. We were not averse to making money, of course, because money fuels the engine of any business venture. But it wasn't the primary motivating factor. So, on every level, this contract sounded like a perfect fit.

★ TIM SHEEHY ★

Until I got to the part about technical specifications and requirements, and I saw the words "135 certificate." This was one of those moments where, if it were a movie, the screen would freeze, and you'd hear the sound of a needle being dragged across a record.

Screeeeeeech!

Whoa, that's not a little thing, I thought to myself. And indeed, it wasn't.

Apparently, in order to fly for the U.S. Forest Service, we needed a Federal Aviation Administration 135 Air Carrier or Operating Certificate. Because there is nothing in the FAA regulations specifically governing aerial firefighting, many of the requirements and certificates stem from applications in other fields. The 135, for example, is basically a charter certificate. So, the pilot carrying organs in an ice chest from one medical center to another will be operating under a 135, as will the pilot ferrying wealthy clients on a private jet or helicopter flying from Los Angeles to wine country in Northern California. Neither of these types of jobs is anything like aerial firefighting, yet all three operate under a 135 certificate. It is not so much a certificate of aptitude or education on the part of the pilot as it is a type of business license attached to the aircraft that will be in use.

For Bridger, the requirement of a 135 was a non-starter. There simply wasn't time to go through the application and review process based on the Twin Commander 500 we had in our hangar, nor enough time and money to make the modifications (specific types of GPS and communications systems, for example) necessary to meet the standards required for a 135 certificate. The wheels of bureaucracy moved much too slowly to accommodate the deadline we faced. By the time we finished—if our application was even approved—bids for the summer would have gone out already. And we'd be broke and perhaps out of business. As I would learn in due course, this was a microcosm of the challenges all aerial firefighters face day in and day out. An intense focus on the mission, on safety, and on excellence is constantly obstructed by misguided regulations that even the enforcers usually find pointless, but they never get changed. It's a case study in bureaucratic mission creep—little by little, almost imperceptibly, changes creep into the contract framework that rarely do anything to support the mission or the fight and usually just present more boxes to check for the bureaucrats back in DC. It's a dynamic hated by pilots, mechanics, and contractors—one which government field personnel hate even more.

Now, I don't think John intended to mislead us regarding the 135 certificate requirement. He was a nice guy, clearly sincere about the overture. But he was hardly an aviation expert, as is the case with most forestry personnel. To most laypeople, a plane is a plane, whether it's a 737 passenger jet or a Twin Commander or an F-14. But these are all very different aircraft with completely different capabilities and specifications. And within each subset, there are subtle differences that can influence effectiveness and eligibility for certain roles. Same with the people who fly these machines. If you tell someone that you have a pilot's license, they will assume you can fly anything from a single-engine Cessna to a 777. But that's like saying you're a baseball player because you play in a beer league on the weekend. It's not quite like playing in the major leagues, is it?

This guy saw me at the airport while I was standing next to a sleek, attractive Twin Commander—seemingly no different from the aircraft used by the Forest Service in air attack—and figured he might as well extend an invitation to join the fight. Can't blame him at all. There is no way that he could have known that we lacked the proper certification or the equipment to obtain that certification or the time to get it done. He just needed help.

And I was eager to be of assistance. The more I read about the industry and the more phone calls I made, the more it became clear to me that this was—and I know this sounds a little *new-agey*, if not downright crazy—our destiny. This was our opportunity to pivot the business in a way that would actually allow us to not just survive, but to thrive in a completely different environment. Most importantly, it would allow us to grow without compromising the ethos that had inspired us from the beginning. We would just have a different customer base than we had anticipated. What I've learned since then is that having the agility to pivot—whether it's fifteen degrees or twenty-five degrees, right or left—is one of the most important attributes of any business. That has been our greatest strength. We've gone from two employees to hundreds of employees in just a few short years because we've never become so dogmatic that we are locked in to one thing. We generally understand what our strengths are and what we want to do, but we are flexible and open to the possibility of change—to the *gift* of change, really.

Still, willingness, even when coupled with desperation, will take you only so far.

We had, at the moment, what seemed to be an insurmountable problem. We were out of time and money—a terrible combination.

I reached out to John Agner again to ask if the 135 certificate was an intractable requirement, or one in which there might be some wiggle room. I had spent an entire career figuring out ways to deal creatively with the rigidity of military bureaucracy, where everything moved at a glacial pace and the chain of command was breached only at great personal risk. But sometimes these risks were worth taking. Sometimes you had to push.

For example...

In 2011, I was getting ready to go on my fourth deployment as a SEAL, this time as a team leader. Things were beginning to wind down a bit in Iraq, but in Afghanistan, we were still ramping up from a military perspective—still chasing Osama bin Laden and playing whack-a-mole with Al Qaeda. It was an extremely busy and dangerous time to be fighting in Afghanistan, and as a team leader, I wanted to make sure that my guys had the best equipment and, subsequently, the best chance for survival. I wanted them to be appropriately outfitted for the job they were assigned to. SEALs, it should be noted, generally have better access to both combat and training resources than regular military teams, but we were on such a hamster wheel of fighting that even Special Forces sometimes found themselves wanting.

A particular point of concern was body armor. By this time, the old-school Kevlar that had dominated a decade earlier was gone. Kevlar was good stuff, saved a lot of lives, but it was cumbersome, promoting fatigue and leading to injuries. By the mid-2000s, Kevlar had given way to ceramic plates that were lighter, easier to insert and carry, and generally just as protective as Kevlar. Most soldiers viewed them as a miraculous advancement in battlefield technology. But they were not perfect. While Kevlar had an operational life that almost always outlasted the service of the person to whom it was assigned, ceramic body armor had an expiration date. Well, not officially, but we all knew that after a while, ceramic plates could develop imperceptible weak points, leading to tiny cracks and fissures that could compromise ballistic properties and rupture at inopportune times—like when an IED blew up nearby, when you took a bullet to the chest, or even simply when parachuting, which can be a much rougher experience than most people realize. To be clear, faulty armor wasn't a common occurrence, but it did happen. Like any new technology, the ceramic plates were an improvement…but perishable. The problem was that you couldn't really be sure of the armor's sturdiness and reliability without complete and thorough inspections on a regular basis. This did not always happen. And as with most types of inspections, a somewhat arbitrary scale was used to determine whether a passing grade was handed out. As a result, troops were sometimes going off on deployment wearing compromised body armor—armor that had been beaten on consistently for a year or two…or more.

As a new team leader, I wanted my guys to receive new ceramic plates while on deployment. Most of them had been wearing the same armor that had been assigned to them on their first deployment.

"We need to have all of the armor inspected before we leave," I told my commanding officer. "Most likely, it will all have to be replaced."

He did not disagree, but as deployment date neared, we were still waiting for equipment to be delivered.

"Don't worry," I was told. "It'll happen."

★ TIM SHEEHY ★

I'd been in the Navy long enough to know that "It'll happen" is hardly a guarantee. It's a brush-off—a reminder that in the military, patience is a requirement. But this wasn't a minor request. It was a matter of life and death, and I wasn't about to sit by idly and wait for the creaky naval machinery to shift into action. In my twenty-five-year-old mind, this was too important to allow the clunky gears of bureaucracy to fix it. As I researched the issue of body armor fatigue, the more I learned, and the more concerned I got. The tiny fractures that occur in ceramic plate body armor pile up relatively quickly. Dropping them on the floor, throwing them off the back of a truck, having them exposed to explosive shockwaves—all these things can degrade the integrity of a body armor plate. Add in all the other stuff a SEAL does to his plates while training, parachuting, diving in corrosive saltwater, climbing walls, and breaching doors and entryways dozens of times, and we were the textbook case of how to fatigue your plates as fast as humanly possible. At the end of a year-long workup cycle, you would deploy to a warzone with the very same body armor plates that you had been beating the shit out of for the past year. And then, if something went wrong, (that is, you were shot through the plate and died) not much would happen with regard to plate examination because at that point, the plate would be totally compromised structurally. In other words, there was nothing left to discuss.

So, I brought it up the chain of command, one person after another, and eventually I was informed that there would be no careful inspection nor, most likely, any new body armor.

"We're just too busy for that right now," I was told.

And I was like, *Wait a minute. Our tactics are rooted in the fact that we know we can rely on our body armor while performing extraordinarily dangerous tasks; that when we enter a room, square shouldered,* front toward enemy, *we are not performing an act of suicide. We are supposed to be presenting a target that is protected by the world's best body armor to anyone who might be armed and dangerous in that room. If they do shoot, they'll most likely hit a plate. Everything we do revolves around our body armor, and you're essentially saying, "We don't have time to fix the body armor?" You've gotta be kidding.*

They weren't kidding. And it was very frustrating to hear it. I felt personally responsible for the people on my team. So I kept complaining, rattling the proverbial cages, and the best I got was, "We'll get you some new armor if we can, but no promises." The implication was that there was some sort of administrative logjam or perhaps some budgetary issues. I don't really know. I didn't think for a second there was any intent to shortchange any units or apathy toward fixing a problem, but I came to understand that the attitude of many in the chain of command was one of reflexive stubbornness and rigidity: "Suck it up; we didn't have body armor ten years ago, and we aren't gonna whine up the chain of command because the gear we have now is no good." There was a valid concern among SOF leadership about appearing overly needy regarding our equipment needs. SOF operators had the most advanced equipment and best quality gear of all the armed forces in the world. Meanwhile, eighteen-year-old Army soldiers were deploying to the same war zone, conducting arguably higher-risk missions, with far less training and capabilities. They weren't complaining about their Walmart-quality gear, but we were complaining about our Gucci gear. It was an understandable position. But no less frustrating for me.

Regardless, the response fell far short of what I considered to be acceptable, so I started looking for "creative" ways to fix the problem and procure plates for my team. If it couldn't be done through traditional military channels—endless requisition orders and formal requests—then I'd find another way. I'd circumvent the chain of command entirely and buy them on the private market. Not that I could afford to buy it myself—outfitting our entire team with new body armor would have cost somewhere in the neighborhood of $30,000. Certainly not a lot of money by military standards, but much more than I could have afforded at the time. But I knew some people.

I started by sending an email to a few folks—people who had access to money and influence—including my brother and my father, explaining what I was trying to do. "We're going on deployment soon," I said, "and we don't have the proper body armor. I need some help here. Any suggestions?"

Less than twenty-four hours after I hit send on that email, I got a call from my second executive officer (XO) on my team.

"Hey, you need to come and see me and the CO," he said. "Right now."

That's weird, I thought, as I hung up on the phone. It did not cross my mind for one second that the email I had sent to my brother and a few friends had anything to do with being summoned by the commanding officer. I had only sent it a day earlier and had kept the chain short and tight. I figured the CO must have had some other reason for wanting to see me.

"Sit down," he said as I entered the room. Then he handed me a hard copy of the email I had sent. My heart fluttered. I was in Special Forces, so I was obviously no stranger to the concept of spying. But it had never occurred to me that anyone was monitoring my emails. Were we spying on ourselves now?

"Do you know how I got this?" the CO said, clearly pissed but trying to hold his temper in check.

"No, sir."

"Well, I got it from the chief of Naval Operations." There was a pause, a deep breath. "And do you know where he got it from?"

"No, sir. I do not."

"He was forwarded it from a friend, who forwarded it from his friend." A sigh. "And do you know who that friend was?"

"No, sir. I have no idea."

"That friend was former President George W. Bush."

I had no answer for that. No response seemed even remotely appropriate. Somehow, in less than a day, a personal email to a couple of family members had landed on the desk of the former president of the United States. The very same president who had initiated the war against terrorism and who presumably would not have been happy to hear that American servicemen might be going off to fight with substandard protective gear. But all I could think at the time was, *Holy shit. My career is over.*

At the age of twenty-five.

I didn't respond with righteous indignation. My motives were pure, no question, but it was also one of those moments where I didn't even try to defend myself. Whatever my intentions, I had violated one of the very core tenets of naval leadership. I had circumvented the chain of command in a way that reflected poorly on every single layer between me and the Pentagon. As I matured in the military and in life, I would come to appreciate this lesson and what it taught me about the realities of bureaucracy. Deep in some room somewhere, there was probably someone giving up holidays and weekends trying to get us better body armor, desperately trying to help the war fighter. And that person was probably going to get a shitty phone call because of me and probably have a lot of really bad days. That made me feel bad. My XO and CO and chain of command were good people working around the clock to get us ready for deployment, and I had made a lot more work for all of them. I was not proud of the impact of my immature actions, and I felt really stupid.

But, in fairness, not every person in my chain of command had held or treated a dead or dying person. I was young, yes, but I had seen a lot of action already. I was so obsessed with trying to protect my guys and bring them home, even if it increased their chances of survival by .01 percent, I didn't care. I was going to do it. So many Americans didn't even give a shit what we were doing day in and day out around the world to help preserve their freedom to do whatever they did every day. The least they could do was spend enough money to make sure my boys had some fucking body armor that worked. I was pissed, and I acted in the best interests of my men. However, the way I did it could have ended my career.

But that's not what happened. Yes, there were repercussions. I was stripped of my team leader designation but allowed to go on deployment as a shooter, a role I knew well and did not mind filling again. More importantly, new body armor was assigned to every team before they went on deployment. I considered that a fair trade: personal career advancement in exchange for the safety of my men. I also had to dress in a hot dog costume and be the dunk-tank dummy at the command Christmas party, where all the children could come along and throw balls at the target and dunk a hot dog in freezing cold January water. Humiliation is a great tactic! But while my XO, Matthew Russell, and CO, Mike Hayes, were swift in their punishment, they also were benevolent and spared me when it could have been much worse.

I learned a few things from that episode, not least the importance of taking a very deep breath and thinking carefully before sending an email. Because once an email is sent, you've lost all control over it. There were things in that email I would have worded differently had I known it was going to end up in the hands of the former president. There was self-righteous verbiage—the kind that comes with youth—that could have been presented differently. But I figured I was just sending a note to some friends and family, and maybe between them, they would come up with the resources to pool cash and buy some body armor for my men. It was not my intention to create a political shitstorm. But that's what can happen with an email. You send it to five people, and they each send it to five people, and those people send it to five people... and before you know it—in less than twenty-four hours, in this case—it reaches someone of immense importance. How much of the email story was true, I will never know. All I do know is this: I sent it, it went bad, I got in trouble, the team eventually got new body armor, and we all came home alive.

In the years since, I grew up and learned how to be a little more cautious and diplomatic, but I always looked for creative ways to circumvent bureaucratic obstacles. Sometimes it was as simple as asking a direct question.

"Any way around the 135?" I asked the gentleman from the Forest Service.

★ MUDSLINGERS ★

"I'm afraid not."

And…sometimes it was not as simple as asking a direct question.

Apparently at an impasse, we struggled to come up with a solution. Shit, I had been a SEAL. Work the problem, I had always been taught. Work the problem, and then work it some more. Don't panic. Don't quit. Calmly, diligently—work the problem. Open doors that appear to be locked. Find an answer. It's out there. Just keep looking.

There is an evolution in BUDS called underwater knot tying. It's a simple task: dive to the bottom of a twenty-foot pool and, while on a breath-hold and under the watchful supervision of an instructor, tie a series of five knots around a pipe along the bottom of the pool with a short piece of rope. The point is, remain calm even while your lungs are burning and surging for air, smoothly follow the procedure by "feel" since you're not wearing a mask or goggles, and tie the knots assigned. Aside from the standard lesson therein, which is to learn to tie knots in any circumstance, there is also the tactical lesson of remaining calm and thoughtful under pressure so you are actually solving problems instead of reacting blindly, frantically.

There was, as it turned out, a way out of this logistical nightmare. It would actually be more expensive than outfitting our Twin Commander with all the equipment required for 135 certification, but it would be significantly faster.

The answer?

Buy a new plane.

Well, two planes, actually.

Let me explain.

Since a 135 certificate is a business license, it's understandable that just about anyone who possesses a 135 will also own an aircraft—the two go hand in hand, just like a taxicab medallion in New York City. You can't own and operate a cab in New York without a medallion. And you'd never have a medallion if you didn't own a cab. For a while, in the days before Uber, Lyft, and other ride sharing services, taxi medallions were worth as much as seven figures on the open market. They were passed down through generations. Medallions were available on the open market—for a price. If someone wanted to break a family's stranglehold on a medallion, it was going to cost them a lot of money.

Similarly, while one path to obtaining a 135 certificate was to outfit our Twin Commander with all the FAA-required equipment and then go through the application process, it was not the only path. We could buy a plane that already had a 135 certification—the certificate and the plane were inextricably linked. We couldn't simply buy the certificate and apply it to our business; each plane used for aerial firefighting needed the certificate. It was a package deal. Take it or leave it.

I started looking around, and I quickly discovered that while buying a plane with a 135 certificate was the easiest and most expedient way to be ready for the bid process when Forest Service contracts went out in the spring and summer, it was also the most expensive. Like, prohibitively expensive. Even the very best deal—the one that was closest to being in our price range—was almost laughably expensive. Since then, Bridger has bought a wide range of significantly more expensive equipment, but everything is relative, right? In the winter of 2015, we were on the verge of bankruptcy, still taxiing down an apparently endless runway. This calls to mind another term in the world of startups: "burn rate." The burn rate is the amount of money a company spends each month—metaphorically setting fire to cash that has been invested—before it begins to turn a profit. Our burn rate was such that by now, we were basically out of fuel. We could not afford the $100,000 to update the Twin Commander; how in the hell were we going to pay double that amount—or *triple*!—for a plane that came with a 135 certificate?

I had no answers, only questions. But learning to ask the right questions is often better than being served up an easy answer. So I started asking around…

The first thing I did was contact an aviation broker. Bruce was the same guy who had brokered the sale of my first Commander and would end up selling us some more. He was well known in the used Twin Commander circles, a nice and knowledgeable guy, but as with any used airplane salesman, sometimes the comparisons with used car salesmen are not too far off the mark. Regardless, Bruce delivered for us every time and knew the world of Commanders better than any other broker out there.

"Bruce, here's the deal," I explained. "I need a 135 certificate, and I need it fast."

He didn't even hesitate.

"I have a client in Iowa. He's trying to sell his whole business: two planes and a 135."

Two planes. I could hear the cash register ringing. "Just so you know, I'm not a rich man," I said.

"That's okay. This guy is motivated. Let me give him a call and I'll get right back to you."

"Bruce?"

"Yeah?"

"What's he asking?"

A long pause.

"Sit tight, I'll call you right back."

That was not an answer. That was, as we might say in the military, an evasive maneuver.

An hour or two later, Bruce called again.

"Okay, I just spoke with Doug."

"Doug?"

Yeah, he's the guy with the 135. Out of Cedar Rapids. Like I said, the whole company is on the market, and he's not breaking it up. And he wants five hundred for it."

"Five hundred…thousand?" I repeated, filling in the blank for Bruce. "Half a million?"

"Yeah, that's the bad news, I guess. Although it's not a bad deal for what you're getting."

I laughed. "Okay, so what's the good news?"

"I think you're the only buyer right now. He said there's been some interest, but nothing serious. And like I said, he wants out."

★ TIM SHEEHY ★

There's an old saying in the recreational watercraft world: the two happiest days in a boat owner's life are the day he buys his boat and the day he sells it. This sentiment is easily transferrable to the aviation industry, with a much higher degree of both satisfaction and disappointment. It takes a lot of money to get started and to chase your dreams, and even more money to keep those dreams alive. Most people eventually cut bait and run. I didn't know the personal circumstances of Doug from Iowa, and I really didn't care. He had a product that could keep Bridger Aerospace afloat. That was all that mattered.

"All right, tell him to send all the info," I said. "I'll look it over and then we'll see. I don't want to waste everyone's time."

I'm not sure what I expected to find in the spec sheet—maybe some piece of information suggesting that five hundred thousand was merely the starting point, that we'd have to put in a bunch of equipment and make expensive modifications in order to meet Forest Service requirements beyond the 135 certificate. I didn't even know where I was going to come up with a half million dollars, so any increase to that figure would have just made the deal more prohibitive. I read the info slowly and carefully. Indeed, Doug from Cedar Rapids had two good-looking Twin Commanders, very much like ours, along with a 135 certificate. But there was something else—something that almost took my breath away.

Doug's business was already on a U.S. Forest Service aerial supervision contract, meaning the Twin Commanders had already been flying air attack on wildfire missions. They were completely outfitted with everything that would be required for the new owner to bid on a Forest Service contract. It seemed almost too good to be true. More than that, actually—it was like the stars aligning, sending us some sort of message about what the company was meant to do.

Given all of that, Doug's asking price wasn't unreasonable. He figured $150,000 for each of the planes, and the rest for the 135 certificate plus a bunch of spare parts and other equipment. I had considered the possibility of negotiating a smaller deal—maybe just one of the planes

OMNR Curtis HS-2L flying boat. Circa 1920. With copious lakes around Canada, waterborne aircraft made more sense than land-based birds. Courtesy of the Canadian Bushplane Heritage Centre.

The Ontario Ministry of Natural Resources is one of the premier aerial firefighting organizations in the world, and one of the top operators of the CL415 family of aircraft. But its origins are far more humble than that. DH60 Gipsy Moth reg. G-CAOX and a Hamilton aircraft in front of the original OPAS hangar, Sault Ste. Marie. The man on the left is unidentified while the man on the right is OPAS Pilot R.F. "Dick" Overbury. Circa 1927–36. Courtesy of the Canadian Bushplane Heritage Centre.

One of the earliest "scoopers," a De Havilland Otter with "roll tanks," literally cylinders without a top cut out slot on them that "roll" and dump the water on fires. Courtesy of the Canadian Bushplane Heritage Centre.

The Otter dropping its load. Courtesy of the Canadian Bushplane Heritage Centre.

One of the earliest Canadair CL-215s, these would eventually become the most widely used and accepted aerial firefighting platform in the world. Courtesy of the Canadian Bushplane Heritage Centre.

My Afghan Interpreter Fida and I in Afghanistan. I would work for years to bring him to the U.S. so he and his family would be safe. It was an incredibly emotionally trying experience trying to save him and others from the horrors of the Taliban.

Golden Shovel! Me, Lee Dingman, County Commissioner Steve White, head of Belgrade Chamber of Commerce, Bozeman City Director Brit Fontenot, David Fine at the ground breaking of our first facility, the combined Bridger/Ascent Vision facility at Bozeman Airport.

Me in the cockpit of our first scooper in production at the Viking (later De Havilland) factory in Canada.

Captain Tim Langton flying Scooper 281 on final for a drop on a fire outside Elko, Nevada, our first ever fire with the scoopers. Paul Evans (not pictured) was right seat on this momentous dispatch! I was jump seat running radios.

Captain Jimmy Stewart flying Scooper 283 around a Pyrocumulus over a fire outside Eugene, Oregon. I was the distinguished photographer (copilot) on this picture.

A steep bank away from a fire after a drop.

Me climbing back into the fuselage of a scooper after we landed on Lake Topaz outside Minden, Nevada, to troubleshoot a landing gear indication. Underwater airplane maintenance was never something I thought I would be doing!

Steve Taylor replacing a commander engine in the field. Steve is our top hand and the spirit of our company. I am privileged to know him. He also serves in the important capacity of godfather to my youngest son, Walter.

After a busy day of three different fires! The three Tims, Landing after dark. Tim Langton, Tim Cherwin, Josh St. Cyr, and me crouching in front.

You will see the angle of the crash appears unsurvivable. To this day, I have no idea why God, or the Universe, or whatever higher power you may believe in saved me and not Jim. After all my near death experiences, this one still confounds me and inspires me to live every day in service of others.

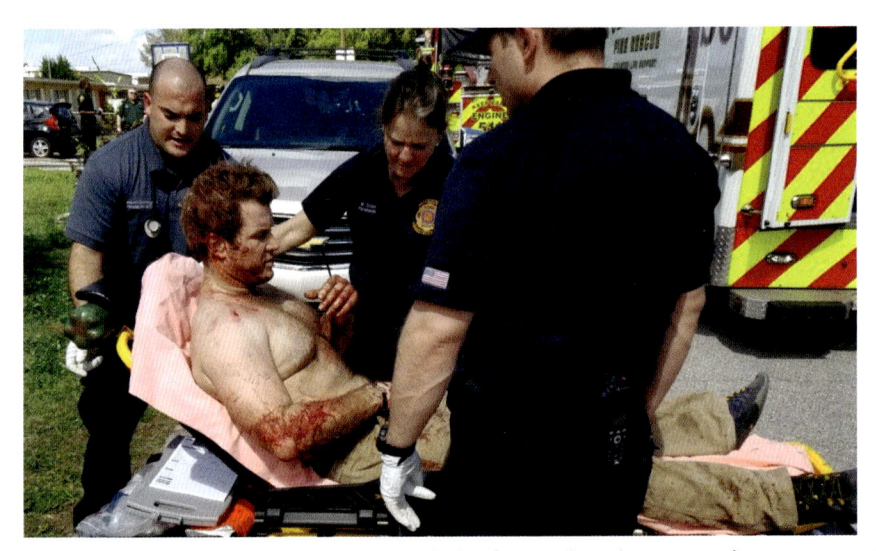

After doing what little I could to assist with the first-aid and post-crash mitigation, I was eventually hauled off to the hospital. Remember, if you're a pilot and survivor of a crash, you are the first first responder; you will be on your own for minutes or hours until the cavalry arrives. I was lucky that it was only about nine minutes, but even after they arrive you will probably be the only person at the crash site that knows anything about airplanes or aviation. Your expertise can make a big difference between life or death (shutting fuel off, preventing explosions, how to open old and complex emergency exits). People crave leadership and direction in a crisis, and you may be able to save lives with your knowledge.

View of a rapidly moving fire line in Eastern Washington from the scooper cockpit.

Me at Bridger Aerospace Base at Bozeman Yellowstone Airport (KBZN) watching the planes head out on their first dispatch in fire season 2021.

The founding team in fall 2014: Sam Beck, Tim Cherwin, my brother Matt Sheehy, and Kevin McDonnell. This is after our first successful test flight of our equipment. Sam and Tim were in the trenches every day. Matt and Kevin helped me keep my head out of the trenches and on mission.

Our scooper after supporting firefighters in Northern Minnesota.

Me and Jimmy Stewart in the cockpit together fighting a fire in Idaho.

You can see me flying the scooper off the Snake River near Lewiston, Idaho. If you look closely, you see this scooper is named after my wife, Carmen, who is the superhero behind everything I have managed to accomplish. A $30mm anniversary present...bought me a few years of gifts!

Me and Jimmy scooping the Snake River in Carmen.

Lake Oroville in CA. You can see how low the reservoirs were getting during the drought.

Me and our top stick, Hank Williams, in Medford, Oregon. I had just flown Hank out some spare parts in the field and always enjoy visiting our fire bases in the summer. Hank has been an amazing member of the Bridger team and it's been a privilege to know him. Behind us you will see the final Erickson DC-7 (another Douglas airframe) that operated out of Medford under a state contract until it was retired the following year.

Me on overwatch on a ridgeline in southeastern Afghanistan with my LPO and top sniper, Matt Pruitt, alongside. I have always been blessed to be surrounded by an amazing team.

Naval Aviation is particularly demanding on the structure of aircraft. That's why airframes build by Grumman and Douglas are oftentimes seen in firefighting—they are known for their naval aircraft and their sturdy construction. A CALFIRE S2 pictured here dropping near Hemet, California, arguably one of the toughest and most effective air tankers. Mario Tama/Getty Images News via Getty Images.

Another Douglas airframe, the MD-87 operated by Erickson Aero dropping on the Holy Fire. MLGXYZ/Moment via Getty Images.

The most common LAT, the BAe 146/RJ-85 family. This RJ-85 is operated by Aeroflite dropping in California. Mario Tama/Getty Images News via Getty Images.

An AT-802 Fire Boss operated by CONAIR. After the closing of the CL-415 production line by Bombardier, Air Tractor filled the demand gap with massive amounts of Fire Bosses around the world. They are now the dominant water scooping platform globally. Alberthep/iStock Unreleased via Getty Images.

The Biggest Kid on the Block. The 10 Tanker, the largest and most affective aerial firefighting tanker in the world. This beast drops at well below max gross weight and can lay down over three times more retardant than the next biggest air tanker. Josh Edelson/AFP via Getty Images.

and a lower price for the 135. But since he had two planes under contract, both impeccably outfitted, it didn't seem worth the hassle. In fact, at this point, we wanted the whole company. For Bridger, which was smack in the middle of a gigantic pivot, it seemed like a smart investment. That we didn't even have the cash to keep one business afloat, let alone invest in another, was almost beside the point.

"Tell him we're interested," I said to Bruce.

"How interested?"

"Just give me a day or two to put something together."

There are moments in the life of a new business where you find yourself doing things you'd rather not do, making phone calls you'd rather not make, and asking for favors you'd rather not ask for. When you're in survival mode, it just happens. And you learn in life to never say never. Up to this point, I had largely financed the business on my own with my wife's and my savings. We were now out of money. But we had a clearer view of what would come next. I didn't have half a million dollars. I didn't have anything except for some equity in the Twin Commander, which I obviously could not—and would not—sell. I had nothing to leverage. For the CEO of a company, I also had limited knowledge of finance and business, a weakness I readily acknowledged then and now. That's why we have a lot of smart businesspeople working for us at Bridger—to handle the financial engineering and structuring to help a growing business attract healthy financing and not make toxic deals that catch up to it later on. In early 2015, however, we didn't have that luxury. We were doing everything on our own. And I needed help.

As I had when I first started the business, I turned to my family.

But first, I called Doug personally to put in a formal offer, rather than going through the broker.

"Look, I'm not going to try to talk you down on price," I said. "You're asking five hundred, then five hundred it is. But I need a transition period."

"How long?"

"How about three years?"

"Deal."

"I'll pay you out over that time, and you help me get everything transitioned to Montana and under my management."

"That sounds good, let's get it done."

With the broad strokes in place—first-year payment up front and then quarterly payments thereafter until the entire five hundred thousand was paid off—I reached out to my brother and my father.

Now, these were not easy calls to make, and neither my brother nor my father initially responded with resounding positivity. It was more, like, "Geez, Tim, we just helped you out with some cash last year (although it was largely my own that got things rolling), and now you need more…and you've decided to become a firefighting company?"

Well, yeah.

It was a lot to ask, and I would have expressed the same amount of healthy skepticism had I been in their shoes, but I was convinced that what must have seemed like an abrupt turn—if not an act of sheer desperation—was the right move to make from a business and philosophical standpoint. For months, I had been taking meetings and pitching our product. I had heard the word "no" a hundred different ways. After a while, you get to know what "no" sounds like before you even hear it. This was different. I knew that if we could come up with the money and the required certification, we would be in a very good position to get a contract with the Forest Service. It just felt…*right*.

It was like diving when I was in the SEALs. People get the mistaken impression that you can see everything clearly when you're underwater because that's the way it's always portrayed in the movies or on their PADI open-water certification dive in the Bahamas. But in real-world diving, it's far murkier. More often than not, it's dark, cold, and silty water. You can't see anything. Hell, you can barely see your hand in front of your face. You're navigating by *feel*—by instinct, training, planning, and experience.

Most of our diving training was done at night, in a muddy harbor somewhere. The whole point was to infiltrate an area without being detected, so you weren't going to waltz under a sunlit sky in the middle of the day or be armed with giant head lamps to make everything brighter and easier to see. Instead, you'd glide through the black water quietly, blindly, trying to find the stanchion of a pier with your gloved hands, slowly feeling around, trying to find the target, whatever it might be, thinking you're lost, thinking you have no shot at completing the mission or the exercise, when suddenly—*boom*! You reach out and grab the right spot on the pier or the keel of the boat. Sometimes the assignment was to disable the propeller of a ship, so you'd swim around for hours in the darkness, trying to find just the right spot to place a charge, wondering if you'd ever find it. And when you did, it was the most amazing feeling—like recovering the proverbial needle in a haystack. Except it wasn't about luck. It was about time and training. It was about patience. It was about knowing that something was right, about feeling it in your gut even if you couldn't see it.

That's the way it felt with the Forest Service contract—an unmistakable feeling in my gut that something was meant to be—that I was on the true and right path. Bridger Aerospace had been in existence for only six months, but I could tell that aerial firefighting—a term I still understood on only a superficial level—was going to be the future of our company. I know it sounds melodramatic, but it's true. We had found our mission, our destiny: fighting fires.

That was the point I tried to get across to my father and my brother as I went through the humbling process of asking for more money. I was lucky. My dad had been a truck driver years earlier before starting his own business, so while he might not have known anything about the aviation industry, he did understand what it was like to take a big swing. He backed me, financially and emotionally, without expecting anything in return. My brother was a bit more discerning. He wanted access to the company's financial records and an equity stake in the business in exchange for a significant cash investment. His diligence in structuring the deal would become the formula that he and I followed for years to come—a formula I often compared to Mulder and Scully from *The*

X-Files, my favorite childhood show. Mulder is always dreaming and thinking of the crazy plan, and Scully comes in and asks the tough questions to make sure they aren't doing dumb shit.

"Fair enough," I said.

We closed the deal by the end of February and put in a bid with the Forest Service a few weeks later.

And then we waited.

CHAPTER 7

FIGHTING FIRE

April 2015

Working on contract for the Forest Service, I realized, was almost like being in a militia. We were part of a vast and loosely connected group of entities ostensibly united in one cause: fighting wildfires. But we were spread out all over the country, answering to any one of a dozen different agencies or municipalities, and you never knew when the call was going to come.

We were still waiting for the first assignment when I was at a trade show in Atlanta. One corollary to the Bridger Aerospace story was the development of our camera and sensor technology that powered our surveillance offerings. Sam Beck had spearheaded a partnership with a small aerial camera company out of Australia called UAV Vision. We bought one of their systems and outfitted it into our first plane, building the rudimentary skeleton of the type of ISR (intelligence, surveillance, reconnaissance) aircraft we had used overseas in the Navy. It was, in every way, the poor man's version of a surveillance aircraft. Looking back, what we had built was ugly, almost farcical. But seeing how fast we developed that technology—putting in hours of careful work every day—makes me proud of our team even today. Mike Bailey, Tom Loveard, and Rob Grew were our partners down under. When we first met UAV Vision, it was solely as a customer. Mike, the founder, was a hilariously sarcastic genius who you either loved (like I did) or hated (like most other people did). His talent was undeniable, and I could see that he was a diamond in the rough. A man with a capability that, when matched with what we were doing, could create an amazing system.

That story is a tale all its own, but suffice to say, eight years later, over four hundred jobs, $300 million in revenue, and $1 billion in enterprise value was created as a result of a crazy British man living in Australia, a surplus NOAA plane, a handful of American vets, and a hell of a lot of creativity. But in 2015, as the adoption of the plane services was slow and we had started our pivot to firefighting, we realized that the tech we had created with Mike could be sold on its own as a product within the U.S. market. So, in yet another turn of businessman's fate, I put on my sales hat and started selling cameras.

For most of the previous year, trade shows had become one of the more depressing parts of my life—a necessary exercise in handshaking and salesmanship that rarely resulted in new business. A few weeks earlier, Sam and I had attended a show in Phoenix, where the temperature was already in the nineties. We had been a last-minute entrant, filling a spot on the outer orbit of the convention hall where traffic was light and the exhibition booths were, shall we say, humble. Certainly, that was the case with Bridger. Sam and I had traveled to Phoenix with a couple of gigantic duffel bags packed with PVC piping and other materials that we used to set up a makeshift booth. We didn't even have a parking pass for the convention center, so we had to drag the stuff for a mile through city streets. By the time we arrived, our suits were completely soaked with sweat. Then we spent two days trying to drum up business.

What type of business? Well, we had three planes now and a 135 certificate that carried with it an active contract with the U.S. Forest Service. We expected a revenue stream at some point; we just didn't know when. In the meantime, Sam had overseen the development of some promising infrared camera technology that we planned to use in our Twin Commanders. We also continued to look for other clients who might be interested in that technology, which we figured could be applied to all kinds of surveillance equipment.

Basically, we were doing everything we could to try to make money and keep Bridger afloat. "Bad times" for businesses can also be the best times for businesses because the bad times force you to be creative, to grow and make money, and to do whatever you have to do to keep your dream alive. Think of it like this: when you're standing at a buffet table, surrounded by more food than you can possibly consume, you don't really think about where the food came from. It's there, so you eat. Comfortably, easily.

But if you're in the middle of the desert with no food and no water, you become deeply concerned with figuring out how you're going to feed yourself, how you're going to stay alive. For most of Bridger Aerospace's first year, we had no money, no business, no customers. We had a lot of leads that became dead ends. But in the process of searching for customers and pitching our services, we had a hell of a time finding a camera that would meet our needs. So, we partnered with UAV to develop our own technology and tried to sell it under the banner of what would eventually be known as Ascent Vision Technologies. *Maybe,* we thought, *we can make some money off this and keep the lights on while we wait for the Forest Service assignments to roll in.*

That was my primary focus at a subsequent trade show in Atlanta: to drum up business for our infrared camera technology and aerial surveillance capabilities and to continue trying to educate myself on all aspects of the aerial surveillance and firefighting industries. I was a neophyte, and the fact that we now had three planes and a 135 certificate didn't change the reality of our situation: we were still out on the long startup runway, waiting to take flight.

I was standing at the humble Bridger Aerospace exhibition booth when my cell phone began to vibrate. I looked at the screen. It was a number I didn't recognize calling from a place where I knew absolutely no one: Hot Springs, Arkansas. I thought for a moment about declining the call or letting it go directly to voicemail, but I decided to answer. We were a new business—you never knew when a potential client was on the other end of the line.

★ TIM SHEEHY ★

"Hello?"

"Hi, is this Tim from Mountain Air?"

It took a moment to process this. Mountain Air, LLC was the name of the Iowa–based company that we had purchased as part of our deal to acquire the Twin Commanders and the 135 Certificate. We were still in the process of integrating them into Bridger.

"Yes, it is? Who's calling?"

"This is Karen from Hot Springs Dispatch. We have a dispatch order for you."

"A dispatch?" I repeated. "What exactly does that mean?" I sort of knew what it meant, in the abstract, but I wasn't quite sure what it meant in that moment while I was in Atlanta at a trade show.

"It means we need one of your planes to go out and fly on a fire," she said.

I swallowed hard as a shot of adrenaline rushed through my body. I hadn't felt anything like that since I'd left the Navy. After almost a year of trying to create a business that worked, to finally have real revenue and real customers was exhilarating!

"Wow," I said, probably sounding less like the CEO of an aviation company than a kid on Christmas morning. I took a deep breath, collected my thoughts. "Okay, give me the details, please."

As it happened, I got this call as I was walking through a parking garage in a cheap suit and tie with my hands full of poorly drafted brochures (I actually have one framed in my office). I had to stop and try to use a crappy hotel pen to write down the information she was giving me on the crappy glossy paper we had. Eventually, I gave up on trying to do that and just wrote it on my hand. Mountain Air's contract was primarily for coverage in the Southeastern United States, but as often happens in the battle against wildfires, they were sometimes dispatched to other regions. In this case, the wildfire was in Texas, and our services were needed. It didn't matter that our planes were sitting in a hangar at Ennis Airport some fifteen hundred miles away.

I hung up the phone and immediately called Tim Cherwin.

"Hey, we just got dispatched," I said. "They need one of our planes to go down to Texas."

There was a long pause.

"We're still training up here."

"Doesn't matter. They need us. We have to go."

"Okay," Tim said, laughing. "I'll get ready."

And that was the beginning—the very first mission aside from spotting wayward cattle. The 2015 fire season turned out to be a rager, the worst in history by a wide margin. Now, of course, every season is a rager, and each season seems to surpass the previous one, but 2015 was something of a watershed, representing the first great leap forward in the escalation of the North American wildfire season. Within a matter of just a few short weeks, all three of our planes were flying regularly, working air attack on fires from Texas to Idaho and all points in between. It was a dizzying, thrilling ride trying to keep up with it all. Along the way, I got to know and love the wildfire community—the people and the businesses—and I was proud to be a part of it.

I should point out that I felt this way even when I was merely on the periphery of the fight. Early in the 2015 fire season, I was not experienced enough to serve as a Forest Service pilot. I had a pilot's license, and I could fly a plane "normally," which is to say, under normal conditions. But flying on wildfires is far from normal, and the Forest Service requires significant specific training and flight time before a pilot is certified to fly on a fire. I could sit in the cockpit and operate sensors and infrared cameras, but I did not have the experience to be a pilot. Which was fine. We had Tim and a couple other guys to handle the piloting duties while I helped out in other ways and continued to accrue enough hours and training to become certified as a firefighting pilot (which would happen by the end of the year), all while trying to run the business.

It was a crazy time. After so many false starts and stalling out on the endless runway, Bridger was finally busier than we'd ever imagined. We had three Twin Commanders out flying every day, fighting fires in

California, Montana, Idaho, Wyoming—wherever we were needed. Our pilots were all over the place. It was busy every single day; there was no rest. And every single day, I had to take out the credit card to pay for fuel, parts, and other travel expenses. And every single day, I would max out one of the cards, switch to another card, and then try to pay off the one that had reached its limit. The problem was that our bank had restricted us to a ridiculously low ceiling on all our cards, like $2,500 to $5,000. And they refused to raise those limits even as our company began to grow and acquire business. To the bank, we were a new company with only a handful of employees and limited cash flow. They wanted documented evidence that we had become a viable and profitable business entity before extending our credit limits. Unfortunately, we needed more credit to keep paying our bills so we could continue to accept assignments, which in turn would bring in more money and allow us to reach the level of profitability that the bank demanded.

This, I believe, is what is known as a catch-22, and it is precisely the sort of backward and restrictive thinking that you find in big bureaucracies (such as the military, which was the inspiration for the Joseph Heller novel that launched the term "catch-22" into the popular lexicon).

But we dealt with it. I ran up the credit card bills as fast I could, and I paid them as soon as the Forest Service money came in. There was often a lag since the Forest Service typically paid its contractors every two weeks, but we survived. We kept hustling and working. Every day was rubber-meets-the-road stuff. In some ways, I felt like a gambler trying to stay one step ahead of the loan sharks and the bookies. It was high stakes poker, but it was worth every penny of the investment.

Not that there weren't moments that demanded introspection—*What the fuck am I doing?* kind of moments. Our second child, a daughter named Evelyn, was born on July 9—Carmen practically delivered her by herself in the hallway of the hospital—and in the days and weeks that followed, I began to wonder about the merits of what must have seemed from the outside to be a rather quixotic adventure.

"We have another mouth to feed now," I said to Carmen, as if this wasn't the most obvious thing. "I just want to make sure that you feel safe and secure about what we're doing here."

Carmen stared right back at me—through me—with a piercing gaze befitting a Marine.

"What are you talking about?"

"I just mean, we're not making any money, and I'm trying to get by on VA disability insurance, and now we have two kids, and…" I paused. "I need to know you're okay with all of this."

"I've never been happier. Don't worry. It'll all work out."

It's funny—in August of that summer, I received a Purple Heart and a Bronze Star with Valor stemming from actions and injuries sustained during a firefight in Afghanistan three years earlier (sometimes it takes a while for commendations to make their way through the pipeline). But I don't remember that firefight being any more frightening than the prospect of not being able to care for my family, and it felt somewhat strange to receive the commendation when our company was fighting for its life. The ceremony itself was a microcosm of the family dynamic at our company. Our congressman at the time, Ryan Zinke (a retired Navy SEAL who would go on to be the Secretary of Interior, which oversees many of our firefighting operations) was the presenter of my decorations. I was aware this was to be the structure of the ceremony and therefore wanted to keep a low profile at work that day. By that time, we had about fifteen employees between pilots, mechanics, and engineers, and I didn't want to distract everyone from their work for an event that honored me. So, I told our office manager at the time, Mariah Simmonds, that I had to go to Helena for a meeting and would be back in the office the following morning. Mariah had come to work at Bridger because her husband, Andy Simmonds, was a British SEAL (Special Boat Service, as they call it) who I had met in the Navy. Small world. They ended up meeting and wanting to move back to her hometown of Bozeman.

★ TIM SHEEHY ★

I was struck by the number of people at the event—a few hundred—and I was even more surprised to find out it was all just for my awards. I was deeply honored and honestly brought to tears by the turnout of all these complete strangers. What really floored me, though, was as I walked onto the stage to have the medals pinned on, the presiding officer was Sam, who was still in the Navy Reserves, and the entire company had immediately piled in cars and driven up to the event. It was an amazing gesture on their part and something that would foretell the family dynamic we tried to encourage in our company.

★ ★ ★

That first summer, we managed to keep all three planes flying safely and successfully despite having only one full-time mechanic on the company payroll. Steve Taylor, who remains with Bridger to this day, didn't move to Montana because he wanted to fight wildfires or because he wanted to get rich. He was simply looking for a better life. It's probably best to let him tell the story in his own words:

"I grew up in Oklahoma, in the panhandle. It was dry, dusty. Dismal. I couldn't wait to get out. Graduated in May of 1997 and enlisted in the Air Force right away. I'd always had an interest in flying and figured this would give me a chance to explore that…and get out of Oklahoma. Funny thing is, my wife is from Oklahoma, too, but we didn't meet until we were both in the military and stationed overseas. In 2007, when it was time to get out, we started talking about where we wanted to settle. We had some friends in Montana and had visited a few times over the years, starting with a wedding in 2001. We really hadn't even given much thought to where we were going to live—we were just focused on detaching from the Air Force and packing up our home in Germany. When the transportation office asked where we wanted to have our possessions shipped, I turned to my wife and said, 'How about Montana?' She kind of shrugged and laughed and said, 'Sure, why not?'

★ MUDSLINGERS ★

"So, we came here in March of 2007. We bought a house in Pony, which turned out to be a pretty good investment and a great place to live, which is all we wanted. We had traveled the world, and we both wanted to settle someplace where we really wanted to live, and then we'd figure out how to make it work. We didn't have any plans beyond that. I mean, there was a lot of pinching pennies to put fuel in the truck the first few years, but we figured it out. I worked at a concrete factory in Montana. I worked on airplanes in California and Georgia. I went to Canada for a while and worked as a welder building bear traps. I did private contract work for some different companies supporting the military, which took me to Iraq and Afghanistan for the better part of five years. But throughout all of this, Montana was my home.

"In Afghanistan, I met Tim Cherwin—we were actually stationed at the same base for a while. At some point after he got out, we started taking about this new company he was involved in, and he asked me if I might want to work for them. I was eager to get out of Afghanistan by then—I wanted to come home and sleep in my own bed. So, I met with Tim Cherwin and Tim Sheehy. I didn't know anything about wildfires, but I knew how to work on planes, and the job was just down the road in Bozeman. They had big plans—they kept talking about doing great things, important things, and it seemed funny since they were just this little company with four or five people working for them at the time. But there was something really appealing about it. I had to take a pay cut to accept the job, but I didn't mind. I'd saved up some money working overseas. It sounds weird, but I had a premonition that the company might turn into something big. I just didn't think it would happen as fast as it did.

"I got to Bozeman just as we were getting those two planes out of Iowa, from Mountain Air, LLC. We had three planes in one hangar, and we just started digging into the federal regulations. Every day it was like, 'Okay, what do we have to do to be legal? What do we have to do to make sure everyone is safe?' It was chaos almost from the beginning, but in a good way. And it's been chaos ever since. But that's okay, because if there's no chaos, then you're not getting better. You're not moving or growing. You're not challenging yourself to support the effort—the company effort and the firefighting effort—by putting out the best possible product."

Firefighters chase the fires; mechanics chase the firefighters. That's just the way it works. Sometimes our planes fly from Bozeman, but more commonly, crews are dispatched to a specific fire base and set up camp there for a period of time—maybe a few days, maybe several weeks or longer—while waiting to be called out to fight a fire in the region. Naturally, those planes need to be properly serviced and maintained while in the field, which means mechanics must be dispatched along with the pilots and planes.

Steve was insanely busy that first year trying to keep our equipment running smoothly and our pilots safe. The first time I ever experienced a wildfire up close and personal—at ground level—was while I was driving from Bozeman to the Lewiston-Nez Perce County Regional Airport in Lewiston, Idaho, in the summer of 2015. I was still trying to accumulate enough hours and training to get certified as an air attack pilot, so I couldn't fly, but there were other ways to pitch in. Steve called one day to say he needed some parts for the plane we had stationed in Lewiston. There were dozens of fires burning in the Northwest and Mountain West at that time, but we were primarily assigned to the mushrooming Clearwater Complex Fire, which had already torched nearly fifty thousand acres near Coeur d'Alene, Idaho. Steve couldn't get the parts he needed locally, and we had to keep our planes in the air, so I said I'd drive them up. It was a 450-mile trek through the mountains—through an active fire zone—but it was really the only way to ensure that the parts would be delivered. And I wanted to do it. I wanted to be of service in a way that took me out of the office and into the fight. I wanted to be out there with our pilots and mechanics.

I gathered up the equipment that Steve requested and threw it into the bed of my 1998 Chevy Tahoe, a big old SUV with a couple hundred thousand miles on it that I'd been driving since high school. This vehicle, passed down from my dad, had taken me all over the country. I learned to drive in that car, lost my virginity in the back seat, and hit a deer late one night on my way home. I still have that car to this day, and I can't imagine ever giving it up. It's almost like a member of the family.

There is no direct route from Bozeman to Lewiston. Most commonly, you drive Northwest to Coeur d'Alene on Interstate 90, then turn sharply south toward Lewiston, which sits on the Snake River. Alternately, if you prefer a more scenic drive, you exit I-90 around Missoula, Montana, and cross into Idaho via the Lolo Pass, a mountainous region in the Bitterroot Range of the northern Rockies. It's a shorter route along serpentine two-lane mountain roads at nearly six thousand feet of elevation. Chains are required in the winter, common sense in the summer. During fire season, well, you just never know.

Having no firsthand experience with wildfires, I was blown away by what I saw as I drove through the Lolo Pass: Fire on both sides of the road, stretching from the ground to the sky. Wind-driven flames whipping through the brush and grass like a lawnmower. Pine trees bursting into flame, the sap igniting like an accelerant and shooting sparks into the air. Trees burning like giant tiki torches, appearing as though they might topple and fall onto the road at any moment.

I'd seen my fair share of shit on deployment in Afghanistan—explosions, firefights, bombings—but this was as potent and awe-inspiring a display of power as I had ever encountered. This was not an *enemy*—not in any sense of the word as I knew it. This was not the result of human malevolence or engineering. This was nature flexing its muscle—an enigmatic, unstoppable force tearing through the mountains with utter disregard for anything in its path. I stared at it in wonder.

Holy hell...this is unbelievable!

With smoke darkening the sky and ash falling like snow on my windshield, I flipped on the headlights and drove slowly down this two-lane highway, both hands gripping the wheel, trying to keep my eyes on the road, but constantly looking off to the side, mesmerized by the strength of the fire, almost hypnotized by its sheer destructive energy. Hearing about wildfires is one thing. Seeing video clips or even witnessing one from a fire tower or while flying air attack high above the flames can give you some sense of the magnitude of a fire. But there is nothing quite like being on the ground with fire raging all around you. Flying in a scooper or a small tanker low enough that you can feel the heat rising up and scraping the belly of your plane comes close to providing that same visceral experience (and an element of danger that, statistically, surpasses all others in the firefighting realm), but I had no frame of reference at the time. It was all so new. To see an active fire burning just outside my car window—to watch the devastation unfold—was...amazing.

Ahead in the distance, I could see something—just shadows at first, shapeless figures moving through the smoke. As I drew near, I could see bright yellow shirts and helmets with headlights, the distinctive uniform of a Forest Service crew. There was at least a dozen of them walking down a hillside in a line, carrying shovels, chainsaws, and of course, the ever-present Pulaski tools. I didn't know where they were coming from or where they were headed, but they had the unmistakable look of exhausted young men on a mission. I recognized that look—the body language and the expression, the hollow-eyed stare. I slowed down as I passed them, tried to take it all in. One of the guys looked up at me as I drove by; we locked eyes. I could see the soot and ash on his cheeks, the utter fatigue in his eyes. I had seen the same look in Afghanistan and Iraq a hundred times, the look of guys who are tired and beat down, but at the same time, in some weird and almost inexplicable way...*happy*. Happy to be part of something important—something they believe in. It's a look that transcends culture, language, and geography. Troops have had that look since ancient times and still wear it today. It's a look that says they are grateful to be part of something bigger than themselves—a team.

And the first thought that went through my mind was, *I am proud to be part of this.*

I could tell by looking at that kid—most of the ground crews in firefighting are young males, just as most of the people who experience combat in the military are barely out of adolescence—that as much as it sucked, he wouldn't have wanted to be anywhere else at that moment. I knew what he was feeling because I had felt it myself. And I wanted to feel it again. And that was it, the point driven home with clarity: this was the type of mission I wanted to be part of. It was what I loved about the military. I hated the bureaucracy and the structure and all the other soul-sucking crap, but I loved the camaraderie of a bunch of young men on the same team on the same mission. Taking care of each other, watching out for each other, and united in purpose regardless of background. There's an immediate feeling of solidarity when you pull into a new base almost anywhere in the world and see someone wearing the same uniform—it doesn't matter that you don't know each other. You're in the same country, on the same team, fighting against the same enemy. There is an instant connection.

That's what I felt when I made eye contact with that young firefighter. Granted, he may not have felt it because he had no idea who I was or what, if any, role I had in the effort. But I could feel it, and that was enough to confirm what I had felt for months: that we had made the right choice by pursuing this path. I wanted to be part of this team, this brotherhood, this family.

I needed it. And I wasn't the only one.

"Our priority is to save things," notes Steve. "To save lives, save structures, save livestock. You know? People don't think about that. Sometimes these cattle are bunched up against a fence, scared to death and trying to get away from the fire. I mean, somebody's gotta help them, right? Somebody's gotta save them.

"You could do all sorts of psychological studies on the kinds of people who are drawn to this type of work—firefighting," Steve adds. "But the part I find most interesting is that it goes across the gamut as far as political affiliation goes. In the fire traffic area—the FTA— everybody is on the same page. Nothing matters except getting the job done. Everything is mission focused. That's a very interesting aspect of it. People come together from different walks of life and work toward the same goal. And focus solely on doing the job. You get that esprit de corps—everyone working toward a common goal. In my experience, it's really hard to find that outside the military."

This is not a unique observation, but it is one that has gained traction and acceptance in recent years: the notion that humans gravitate toward tribes with their shared sense of purpose and belonging and that losing the tribal affiliation can be devastating. The writer Sebastian Junger, who has reported extensively from battlefields all over the world, most notably and effectively in Afghanistan, has addressed this subject on multiple occasions, probably because he has seen it—and *felt* it—up close and personal. It helps explain why so many service men and women struggle to integrate effectively back into society when they return home from combat—not merely because of PTSD and traumatic brain injury or other maladies (although those can certainly play a role), but also because, on some level, they miss the experience. Which is not to say they miss being shot at or they miss killing or seeing friends killed—it's a lot more complicated than that. What they miss is the shared sense of purpose, the feeling that they are representing something bigger than themselves, and the bond that grows out of that struggle. It is both a philosophical hunger and a neurological phenomenon. The adrenaline that is repeatedly pumped through the body and the brain during stress, day after day, changes you. And it makes the world that you return to upon completion of military service seem almost unrecognizable.

And lonely.

We have a high percentage of military veterans working for us at Bridger in part because the company was founded by vets and part of our mission is to support those who have served their country, but also because firefighting, I believe, attracts the kind of people who have served in the military. Not because they are looking for a thrill, necessarily, but mostly because they are searching for meaning in the otherwise humdrum existence of work.

★ MUDSLINGERS ★

And it isn't just the veterans who've seen combat who miss it. We have people working for us who were mechanics in the service or who filled other support roles, and they'll be the first to tell you that it gets into your blood. These are not guys out there cracking skulls, getting shot at. They're fixing planes or helicopters, far from the front lines. But they know how important their work is. And they take it very seriously.

"My wife says I'm a desert rat," said Steve, "because I'm always going back in the sandbox."

There is a contact high that comes with just being in proximity to the action and to knowing that you are contributing in some way to the mission. This is true in firefighting, and I found it to be true in the military. This will sound somewhat corny, but I remember my first deployment as a SEAL, in Iraq, and how I excited I was just to be there. I remember running around Baghdad International Airport on one of my first days, trying to get a workout, and as I jogged past the guard tower, one of the soldiers looked down and waved. And I waved back. I didn't know him, but it felt as though we were connected. A few minutes later, a helicopter flew low overhead, on approach. A soldier in the doorway looked out, gave me a little two-finger salute, and I returned the gesture. I was brand new to Iraq, brand new to the base and everyone there, yet we all were connected. We were part of the same team—from the officers down to the mechanics and logistical support staff. We were in lockstep, transported half a world away from friends and family for the same reason, forced by circumstances and proximity to embrace our commonality rather than bristle over our differences. There are no Democrats or Republicans in foxholes, just Americans.

It brought me back to a memory from 2006, when I was still at the Naval Academy but going through Army Airborne School as part of my Ranger training. In jump school, you spend all day walking around in uniform. Everyone looks the same: lean and hungry, shaved head. Soldiers in training. Under those conditions, you form bonds quickly without any consideration of the background of the person next to you. All that matters is that you are both there, working hard, suffering, learning, growing. Shared experience—especially shared misery—can lead to feeling like you're best friends in a matter of days, even though you don't know anything about each other. You eat together, sleep together, train together, jump out of airplanes together. And you talk. This was before the widespread influence of smartphones, so no one was staring down at their phone whenever they had a free minute. Instead, it was like, "Hey, I got a free minute, let's shoot the shit. We're buddies, right?"

On the weekend of our first liberty, when we could leave the base and go out in street clothes, it was jarring to discover that in fact we were *not* brothers from different mothers. We were completely different people, tossed into the same lab experiment and compelled to get along. I grew up in the Midwest in an upper middle class family. I went to the Naval Academy. When I went out on liberty, I wore the uniform consistent with my background and upbringing: khakis and an untucked polo shirt. One of my buddies, who was Mexican, wore long baggy shorts hanging off his butt and a baseball cap turned sideways. Our Black squad member was dressed like he was going to play basketball. Our white Army counterpart was dressed like Joe Dirt without the mullet. We all kind of looked at each other and laughed, and then of course we started busting each other's balls, and I couldn't help but think, *If I ran into these guys on the street, without knowing them, we wouldn't even talk to each other; we would think we had nothing in common, based purely on assumptions related to appearance.* And other people who saw us in town were likely thoroughly confused by us sitting together, probably assuming I was interviewing these guys for a job at the restaurant we were in! But we were four of the top performers in our class. We shared a similar work ethic and a drive to succeed. We had similar senses of humor. We watched out for each other. We were friends. It didn't matter where we came from or how we had arrived in this place. All that mattered was that we were there, sharing a common goal.

That feeling is contagious, and once you've had it, you're always chasing it. I'll be the first to admit I have an addiction. My wife even says it about me: "You have a problem. You always want to be at the center of the action. Whatever is going on, you need to be a part of it. And if you're not, you feel bad about yourself."

She's right, and being a Marine, she understands the behavior, though she'd like it if I throttled back a little. But I don't seek danger so much as I seek situations and experiences that make me feel alive, and by that, I mean contributing to a cause in some way—helping people. I didn't mind Airborne School, but I'm not the kind of person who is going to go base jumping for the pure thrill of it. I do not actively seek pointlessly dangerous experiences. Aerial firefighting is dangerous, no question about it. But it certainly isn't pointless. (Whether you believe combat is pointless is an entirely different discussion and one that is generally too deep and philosophical for these pages; suffice to say that I fall into the camp that believes it is sometimes a necessary horror.) When I was introduced to firefighting, I instantly recognized it as a dangerous pursuit, but also one of deep importance. Its front-line personnel were as committed and connected to the cause as anyone I had met in the military. And all I could think was, *Okay, this is it. I'm back home.*

CHAPTER 8

LET IT BURN

After I drove to Lewiston to deliver parts and supplies to Steve, I spent a few days there, helping out with maintenance and operating sensors for our pilots. This was a project fire, so the days were long and hard. The sun doesn't set until ten o'clock at night during the summer in the Mountain West, and since total darkness doesn't fall until nearly eleven, flight crews were out until the last possible minute every night. Of course, when planes are done flying for the day and an exhausted pilot clocks out, there is still work to be done. Every aircraft must be checked before it is cleared to fly again the next day. Sometimes significant work is needed: oil changes and other normal but time-consuming maintenance and repairs. I would often stay out through the night to help Steve in any way possible. I was not a mechanic, but I knew how to do oil changes, and I could certainly hand him tools and make sure the spotlights were pointed in the right direction. Anything to make his job a little easier and more efficient.

With one exception—a pilot named Fred Stone who joined Bridger when we acquired Mountain Air—all of us were new to the firefighting world. We were sponges, soaking up as much as we could and learning as we went along. One night at the hotel bar, after another impossibly long day, I was talking with Will, one of our pilots. He was a former bush pilot, flying on fires for the first time, and he was as wide-eyed as the rest of us. We were just sharing observations about what an incredible scene it was, what a challenge it was to contain something so ferocious and bent on destruction.

Like any hotel bar in that region during the summer—this was the Red Lion Hotel in Lewiston—the place was full of firefighters. Ground crews, air crews, support staff—wall-to-wall firefighters from a dozen different agencies and private contractors. That's just the way it works in a fire zone. Everyone is there for the same reason. You weren't going to find many tourists at the Red Lion Hotel in the middle of a wildfire. Everyone was talking about the fire, rehashing the highs and lows of another exhausting day.

"Man, it's crazy out there," Will said to me, his tone reflecting a mix of wonder and apprehension. "I saw trees exploding today. Like, literally just bursting into flames and blowing up. I've never seen anything like that."

I nodded in agreement. I still couldn't shake what I'd seen coming through the Lolo Pass. Part of me wondered if this was a typical scene in a big fire or something more extreme. I had no frame of reference.

"Yeah, it's unbelievable," I said.

Suddenly, there was a voice from the end of the bar, a few seats away. A dismissive grunt, clearly directed at us. I leaned forward, saw the guy—maybe mid-forties, kind of weathered, perched over a beer, half looking at us, half ignoring us. I had no ideas who he was.

"Excuse me?" I said.

He smiled disarmingly. "You guys are new, huh?"

I shrugged. "Yeah, I guess."

"Well, trust me, you ain't seen nothing yet."

I tried to stifle a laugh. "Is that so?"

"Yeah, you should have been here in ninety-nine. Fucking rager, that was." He turned to another guy seated next to him. "Remember that one, Matty?"

Matty nodded, smiled. "Now *that* was a fire."

A few other people nodded in agreement, but one guy, who obviously knew Matty and his buddy, responded with disdain.

★ TIM SHEEHY ★

"Ninety-nine? Come on, man. What about Durango, 2002?"

A murmur of approval, respect.

And just like that, we were off to the races, our conversation bleeding into a larger one—a bar full of exhausted and semi-inebriated firefighters playing an impressive game of one-upmanship, name-checking fires from Yakima to Yuma, stretching across decades. The amazing things they saw and did! The tanker pilot nearly losing a wing when he flew too low and got caught in an updrift; the smokejumper who got blown off track and landed next to a wall of flame, separated from his mates; the air attack pilot who lost an engine and barely made it back to base. A verbal ping pong match of escalating intensity. It was like Quint and Hooper comparing battle scars in *Jaws*, with Quint's story about being a crew member on the doomed U.S.S. *Indianapolis* eventually ending the debate.

Will and I had nothing to contribute, not in our first season on the job. Of course, I could always pull out war stories from Iraq or Afghanistan, but in fire circles, I learned quickly that although the firefighters respected your military service, the last thing they wanted to hear was you one-up their fire discussion with an unrelated story from the war just so you could preserve your manhood. It was far more polite and, frankly, you earned more respect if you just listened to their stories without changing the topic to your own experience. In later years, in similar scenarios, I would find myself gently putting my hand on other team members, fresh out of uniform, who hadn't discovered this truth yet.

That night, we just sat there and took it all in, absorbing the atmosphere and marveling at the camaraderie. I was new to the firefighting world, but I'd been through this sort of indoctrination in the military on multiple occasions. There is always a certain amount of awkwardness when you're the new guy in any situation, but for me, there was also a familiarity to it. There were so many long days in the Navy that ended with a bunch of guys sitting around, telling stories—"You think this is something? Hell, I was in Iraq when Bush One ordered the invasion!" It was the same sort of atmosphere—guys bragging and telling stories and even making fun of each other...all in the spirit of collegiality and teamwork. And it confirmed, once again, that I was right where I belonged. Or would be, anyway, with a bit more experience.

But over the course of that first summer, there was something else that I recognized, something also found in the military, that could undercut the camaraderie and closeness as well as the effectiveness of the mission: bureaucracy, thick as sludge, gumming up the gears and impeding progress, rewarding complacency and stagnation and waste while discouraging innovation. It started, as it always does in calcified bureaucracies, at the highest levels—government agencies fighting over power and territory and funding. Inevitably, though, it trickled down like a toxic stream to those on the front lines.

This complex reality was driven home one morning on that same trip, after our planes had taken off for the day. I was hanging out at the base, shooting the breeze with some other guys, talking about how intense the fires seemed to be, just trying to make conversation and contribute to the cause.

"Hopefully we can hammer this thing down quickly and get it under control," I said. Most of the other guys nodded solemnly, but one person, a pilot, kind of straightened up and grunted.

"Well, we don't want it to go *too* fast," he said. "There's a lot of overtime pay to be earned out there! We put it out, it's back on salary!"

I didn't say anything because, again, I was the newest of the newbies, and I was merely trying to absorb information and be of service in whatever way I could. Sometimes you learn more by simply watching and listening than you do by inserting yourself into the conversation—at the very least, wait until you know what you're talking about. This was neither the time nor the place to get into a deep discussion about bureaucratic sclerosis or the wasting of taxpayer dollars. But as I let that statement roll around in my head, I couldn't help but feel a sense of unease.

Is that really the primary concern? Keeping the spigot of funding wide open by any means necessary?

★ TIM SHEEHY ★

This was the first time that I had even considered the possibility that despite the shared sense of purpose so prominent in the firefighting community—which was extraordinary and inspiring—there also existed a troubling undercurrent of complacency, of embracing or at least *accepting* the status quo because, frankly, there was so much money at stake. I've since come to realize that this is not a feeling shared universally, but it does exist, and to deny its existence is to impede the efforts of those who understand the importance of change. Keeping one's head in the sand is rarely a good strategy when it comes to transforming large organizations. Get in the fight or get out of the way.

There are significant sections of the wildfire universe that believe extinguishing fires as quickly as possible is the wrong answer to our problem. Now, there is nuance to this discussion. As noted, and widely accepted, forest management sometimes depends on natural or prescriptive burns that reduce fuel load and, subsequently, the likelihood of a catastrophic project fire. But those are generally in areas with barely a human footprint. And, of course, as with the devastating 2022 Calf Canyon/Hermit's Peak fire in New Mexico, sometimes even the most carefully monitored prescribed burn can get out of control and end up destroying property and threatening lives. That does not mean that prescribed burns should not be utilized, but merely that they are one weapon in the arsenal, to be employed with caution and respect.

This particular firebase conversation, however, smacked less of concern or common sense than it did laziness—or, worse, greed. I wouldn't call it malevolence; anyone who climbs into a plane or picks up a shovel to fight wildfires clearly has a capacity for goodness and a desire to help. That said, even in positions that are demonstrably service-oriented, there is the potential for self-interest, if not outright corruption, leading to a response that is not necessarily in the public's best interest. In the case of wildfire management, it seemed, there were not just power struggles and territoriality among supervising agencies, but an overall sense of ambivalence toward the objective as well.

Maybe we ought to let this thing burn for a while.

★   MUDSLINGERS   ★

I understand that this is an unpleasant subject to discuss, both from the perspective of the homeowner or rancher who could lose his home or the many firefighters who would never place personal financial interests over the well-being of the people they serve (and who would be offended at the very suggestion). But I don't think you can write about the history of aerial firefighting without addressing the matter honestly and transparently. As I came to realize that summer, the truth is that there is an entire ecosystem built around the firefighting industry—from pilots and mechanics and ground crews to the hotels and restaurants that cater to them when a fire breaks out in the vicinity. If there is no fire, there is no money. And the faster that a fire is extinguished, the sooner the money dries up or goes elsewhere.

It might seem ridiculous to worry about a shortage of work to keep the wildfire industry busy given the extraordinary expansion of the season in recent years, not to mention the gnawing sense that firefighters will forever be overmatched against nature. But old beliefs and protocols die hard, and clearly there were some in the industry who saw nothing wrong with milking every fire for what it was worth despite the risks and the blurring of ethical boundaries.

In some ways, it reminded me of the defense industry. There are fewer gray areas with fire, but there are similarities. Vulgar as it might be to admit, destruction—or at least the *threat* of destruction—is good for business, so you don't want to eliminate that threat entirely or too quickly. There is an assumption that eradication of a threat or an enemy (in this case, fire) is the ultimate goal, but whether through mismanagement, mistakes, or deliberate slow playing of a solution, that often is not the case.

Fast forward a bit to the spring of 2022 as I watched with some amusement, but mostly dissatisfaction, as representatives from the U.S. Forest Service and other agencies testified before the U.S. Senate Committee on Energy and Natural Resources about how aggressively the agencies typically responded to wildfires. The obfuscating and falsities (intentional or unintentional) were embarrassing to witness. But such behavior has become almost normal—the undersecretaries trying to assure everyone that wildfire management is ongoing and vigorous; from the initial attack to the snuffing of the last flame, the goal is to stop the spread of any given fire as quickly and safely as possible using every resource at their disposal, while intimating that they have all the resources they need. It's a false narrative—*"Everything is just fine"*—that has been floating around the industry for so long that it tends to provoke nothing so much as collective eye rolling.

Study after study by different independent organizations routinely shows that everything is *not* fine. We are under-resourced and not particularly good at using the resources we do have. Simply put—we could do *everything* so much better. Obviously, to a certain extent, I'm biased, but I'd like to think that I'm a reasonably objective person. I do believe that most of the problems with the wildfire industry stem more from the governmental and bureaucratic side of the equation than from the business side. But as I like to say about the wildland fire community, you will find an amazing bunch of people, dedicated public servants… trapped in a frustrating and often bewildering bureaucracy.

Your local fire department is something you likely take for granted. It's probably a nice brick building that sits on prime real estate not too far from your place of work or home. It is staffed by highly trained men and women who, on the taxpayer's dime, have gone to schools, training courses, certification programs, and recurrent rehearsals to ensure that their skills are honed to save your life on the extremely rare occasion you need them. They drive million-dollar fire engines and wear tens of thousands of dollars of cutting-edge gear, all to ensure that when you dial 911, they are at your door within minutes, ready to save your life, rescue your cat, douse your kitchen fire, or pry you from a burning vehicle.

The history of the modern fire department is an interesting thing, worthy of its own book. To summarize briefly, as urbanization across the United States exploded in the late nineteenth century, structure fires became a scourge that would kill thousands each year and cause massive economic damage and social disruption. As American cities began

★ MUDSLINGERS ★

sprouting up, most of the buildings were of a shoddy wooden construction, built hastily and to the lowest acceptable standard. Few of these were the architectural stone masterpieces of Europe, New York, or Boston. As Americans migrated west, cow towns, trading posts, and railroad spurs were rapidly transformed into metropolises of hundreds of thousands of people, living and working in rickety wooden buildings. As the age of electricity dawned, so too did the increase in urban fires. These were not barn fires in the middle of nowhere or apartment fires in a brick high-rise; these were infernos that would literally jump the road, ignite the next block, and spread—pardon the pun—like wildfire.

As these urban infernos became more common, the traditional "fire brigade" of a handful of volunteers galloping around the city was not going to cut it. With little to no standards regarding building codes, firefighter training, and department resourcing, it was, quite literally, the wild west. Hoses didn't match fittings, professional fire brigades didn't exist or had mismatched gear, and buildings were literal death traps. After catastrophic fires in Chicago, Kansas City, and Boston, it was clear that a change was needed. So, in 1896, the electrical companies that were the largest building contractors at the time formed what would become the NFPA (National Fire Protection Association), whose mission was, "To be a global, self-funded nonprofit organization, devoted to eliminating death, injury, property and economic loss due to fire, electrical and related hazards."

Over the ensuing decades, the NFPA helped define standards throughout the United States, culminating with the NFPA 409 and 1710 code reviews that were adopted across the country starting in the 1970s. Although considered divisive in many communities, citizen deaths by structure fire have dropped more than 50 percent and firefighter deaths more than 60 percent since the adoption of those codes. There have also been immeasurable economic and public health gains. There was no secret weapon or magical technology that made this happen, but rather a disciplined focus on standards, procedures, and innovation, including proper resourcing for aggressive initial attack.

Other critical elements were standardization of construction, training and firefighting procedures, and innovative technology. Sprinklers, geographically distributed fire stations, better gear, respirators, pump trucks, fire hydrants, proper funding, and sufficient staffing ensure that in almost any urban area in the United States today, any structure fire will have a department response within five minutes. Getting to the fire early and with the appropriate resources ensures that citizens can be rescued, fires can be contained before they become unmanageable, and firefighters are not risking their lives with insufficient equipment and resources.

Over the past ten years, as the Wildland Urban Interface has expanded exponentially, more structures have been lost in wildfires than urban structure fires. Whether we like it or not, wildland firefighters are now in the structural firefighting business. As in the nineteenth century, with its rapid urban growth, we now have massive habitation developing in areas prone to wildfire, turning the WUI into the equivalent of a nineteenth-century Chicago tinderbox. We must now look to resource our aerial wildfire response with the same rigor, aggression, and commitment as we do our urban fires. If you told a resident of Dallas that it could be hours or days before a fire engine would arrive at their home, they would find it unacceptable, and a deep desire for change would take hold. But that's what happens sometimes with wildfires, as citizens watch their lives unravel as they wait desperately for the arrival of an air tanker or ground crew.

Letting wilderness fires burn is a land management decision that must be dictated by policy that prioritizes the stewardship of that land. A critical and important piece of that is controlled burns and natural fires. Putting every single fire out immediately all the time isn't the answer. However, we must have the resources to respond quickly and aggressively to those incidents that do require full suppression to protect those in need. Much like our urban fire stations are manned 24/7 with brand new fire engines and the latest equipment, so too should our aerial firefighting apparatus be ready to respond. As the aerial firefighting industry continues to grow, we are in need of an organization to help ensure cross-industry and cross-agency standards, collaboration amongst operators, and aligned messaging with our agency partners to the media, congress, and most importantly, our citizens.

I would imagine that you and your family and neighbors rarely approach your city council and demand cuts to the pay and capabilities of firefighters because they haven't saved your life in the last week. It is generally considered political suicide for a politician to advocate for cuts to local fire departments. Everyone wants to be pro-firefighter. On 9/11, hundreds of firefighters died while running into a towering inferno. It was the profile in courage that Americans expect out of their first responders. And it arguably placed local fire departments on the list of "untouchable" public servants for a long time.

Unfortunately, wildland firefighters, and specifically aerial firefighters, receive no such treatment. The agencies that manage aerial firefighting fleets and companies are in a constant battle to find ways to do it cheaper—to pay its contractors less and pinch every penny while getting the minimum possible capability out of the groups providing this critical first responder capability. Citizens who live in wildfire-prone areas in the West usually have no idea that while their urban counterparts enjoy a lavishly funded and appropriately resourced department, the planes and ground teams that are tasked to respond to wildfires are anything but. Most ground teams are staffed with highly motivated and skilled individuals who make little more than minimum wage and usually have a passion for both the work and the lifestyle. Aerial firefighters are contractors, private employees working for private companies that are contracted with the government. And although there have been honeymoon periods over the years, typically the dynamic between the government and the contractor is one of suspicion, mistrust, and constant cost oversight. Most contracts awarded to aerial firefighting vendors are evaluated under the paradigm of "Lowest Price, Technically Acceptable," which is a euphemism for "Bottom Dollar Wins." Despite the fact that the aircraft needed to do this job are highly specialized and the pilots require years of training on top of decades (usually) of experience to operate in this highly hazardous environment, the government believes that the best way to contract vendors is to find the cheapest ones to do it.

As Mercury Seven astronaut John Glenn aptly said, "As I hurtled through space, one thought kept crossing my mind: every part of this rocket was supplied by the lowest bidder."

This brings into focus the reality of the wildland fire apparatus for almost the entire U.S., save California. Most of the agencies responsible for wildland fire mitigation are "land management agencies." They are not emergency response organizations. It's akin to taking the largest fire department for the largest city in the country, and placing it under the direction, budgeting, and management of the parks and recreation department. That's not meant as a disparaging remark—it is merely a fact. Or imagine if the U.S. Marine Corps, whose job is to rapidly respond to national security needs in high-risk situations, was administered by the Department of Labor. You would have a misalignment of priorities, motivations, and capabilities.

This is the situation that exists today in wildland firefighting in the U.S. Consider the U.S. Department of Agriculture, which owns the U.S. Forest Service, which owns the largest fire department in the world. The USDA is an organization that exists to do a handful of things, but mainly just one thing: regulate and oversee the production, pricing, distribution, and subsidization of U.S. farming produce. The USDA mission statement: "We provide leadership on food, agriculture, natural resources, rural development, nutrition, and related issues based on public policy, the best available science, and effective management."

Now, let's look at the U.S. Forest Service mission statement: "To sustain the health, diversity, and productivity of the Nation's forests and grasslands to meet the needs of present and future generations. For more than 100 years, the Forest Service has brought people and communities together to answer the call of conservation."

How about the Department of Interior Bureau of Land Management? That is the other primary agency responsible for aerial firefighting in the U.S. "The U.S. Department of Interior Bureau of Land Management's mission is to sustain the health, diversity, and productivity of public lands for the use and enjoyment of present and future generations."

You get the point. This could go on for pages as we dissect each agency that has a responsibility for wildland fire mitigation in the United States. Nowhere in the mission of any department, agency, or bureau are the words "fire," "emergency," "first responder," "protect," "serve," or "save" ever mentioned.

How about the Bureau of Indian Affairs mission statement? "The Bureau of Indian Affairs' mission is to enhance the quality of life, to promote economic opportunity, and to carry out the responsibility to protect and improve the trust assets of American Indians, Indian tribes and Alaska Natives."

Now, here we are trending in the right direction. "Protect" is an operative word. It means something—in this case, that the bureau will protect the land and people under its charge. "Protect." Its very definition means: "Keep safe from harm or injury." Words have meaning, and their usage impacts the perception of people both inside and outside an organization; words shape how you behave within that organization, who joins the organization, and how that team molds itself around its mission. I would also argue, from personal experience, that the BIA is unique among federal agencies in that it has an aggressive firefighting posture. When a fire starts on a reservation, the goal is to extinguish that fire—fast! The BIA doesn't dither about letting it burn, or the costs at risk. They just fight the fire, because it's *their* land, they live there, and they are accountable to the citizens who depend on that land for their livelihood. It's no coincidence that the organization that has the word "protect" in its mission is also the one that is generally regarded as the most mission-oriented wildland firefighting agency in the federal government.

Now, let's talk about states. CAL FIRE "provides all hazard emergency—fire, medical, rescue and disaster—response to the public and provides leadership in the protection of life, property and natural resources." That's a mission statement that clearly connotes a commitment to safety, excellence, and clarity around first response. And of course, CAL FIRE is the definition of an airborne fire department. Although many disagree with some of CAL FIRE's choices for aircraft, tactics, contractors, and technology, it's hard to deny that CAL FIRE makes its mission clear and drives to execute on that mission with excellence.

Mission statements matter. They define the purpose of an organization and flow through the manpower structure of that organization by defining the metrics for success, standards for excellence, cultural ethos, and funding priorities of said organization. If an organization's mission is to "Watch flowers grow and properly steward the vegetable garden," one could argue that implicit in that mission is protecting that garden from invasive species, flower-hungry deer, and bulldozers that seek to demolish the garden and turn it into a parking lot. If the mission statement of an organization is "To protect and serve flowers in our garden and harvest as many vegetables as possible annually to feed the village," it is logical to assume that the priorities of the first group would be to purchase gardening gloves, a trowel and a sunhat, and the staff would consist of botanists and ecologists—a worthy staff and a worthy budget for the mission. The latter group would buy, among other things, a fire hose and sufficient fencing.

The people attracted to organizations will self-select, as well; no matter how hard you try to recruit certain people, they will look at your mission statement and the culture of your organization and determine if it works for them. People that want to jump out of planes, take risks, and protect others will probably not be attracted to the USDA. They are much more likely to enter the military. And if you're trying to protect a thousand homes from an aggressive wildfire, the man or woman who joined the military is exactly who you want.

Should the Forest Service change its mission to better focus on firefighting? No. But shouldn't the largest firefighting organization in the nation have a mission statement that in some way enshrines service? Wildfires are not a niche problem for obscure western communities anymore. They are an existential threat to the millions of people living in expanding fire-prone areas and a general health threat even to those who do not. In 2021, an estimated thirty thousand people died of respiratory issues thought to be related to the inhalation of wildfire smoke. For the aerial firefighting community, this paradox in mission and culture manifests itself most notably in two areas: contracting and associated funding, and operational deployment. The contracting process to procure and implement aerial firefighting aircraft is outdated, burdensome, and ineffective. So much so that protests from vendors who spot blatant mistakes in the contracting process have become the bane of the government contract offices. They have successfully, in many regards, painted the vendors as the bad guys who protest the government so they can make more profit. In fact, the vendors are protesting a process that is deeply flawed and not representative of the realities of acquiring multimillion-dollar aircraft, hiring a staff of hundreds, and putting lives at risk to protect people, all while having precisely zero dollars in income guaranteed. No other government agency does business like this, and it's a travesty that the companies and individuals who risk it all to fulfill this important mission are oftentimes treated like disaster profiteers. As I have said to many government officials, "YOU made the decision to contract this function to private industry, not us. If you want us to exist in a no profit, pure cost basis world, then nationalize the industry and absorb our companies and capabilities into a federal agency. Otherwise, accept that the dynamic you have created is one of private company cooperation with agency; and intrinsic in that is the need to have profit."

Furthermore, the manner in which contracts are structured, which is all risk-based to the vendor, usually precludes bank financing on assets. Without guaranteed revenue on a $32 million plane, most banks will not provide financing. So more expensive private financing avenues have to be explored, which makes the acquisition of these aircraft 10 to 20 percent more expensive. This cost is, of course, passed onto the customer, because it has to be. You can't contract someone for a job and expect them to perform their mission at a loss. Yet this seems to be the goal of the U.S. federal agencies—to drive margins for their contractors as low as possible while simultaneously raising the bar for technical requirements.

On top of all this, aerial firefighting is restricted by Department of Interior and USFS rules that require a fire aviation provider to be a "small business" as defined by the Small Business Association. That size standard prevents any large company or financial institution from being involved in the aerial firefighting business. On its surface, maybe that sounds great: only mom-and-pop, family-owned companies allowed. That might have worked in the 1980s, when leftover military aircraft were plentiful and cheap. In the 2020s, when multimillion-dollar new aircraft dominate, an entirely different financial model is required. Aerial firefighting requires the ability to finance assets in a structure that is almost impossible for small companies. You have to take funding at a very high interest rate that is not collateralized by the contracts, because they are all usage-based. Therefore, the ability to provide new and safe assets is further restricted by this paradigm.

The lack of stability and predictability in fire contracts creates a dynamic where the fire contractor, who owns a $10 million aircraft, has to plan to offset not just the cost of the asset but also the cost of overhead for said asset with an unknown usage rate. If you have no guarantee of any work at all, you have to price your asset according to what very likely results in minimal usage. For example, if you have to buy a $40 million plane and hire ten people to maintain, staff, and fly it, those are fixed costs. You are going to carry those costs regardless of whether you fly fires or not. So, whether you fly for 60 days and 100 hours, or 150 days and 400 hours, the costs to you, the owner, are largely the same. Therefore, it becomes a simple equation: it costs me $6 million a year to own and operate a large aerial suppression aircraft, so I have to earn $7 million a year to manage those costs and have enough left over to cover debt—and, hopefully, have some cash leftover for future investment, growth, and profit. Without a contract that gives me *any* guarantee (which is how almost all fire contracts in the U.S. are structured), I have to assume the worst outcome, so that I don't go out of business. I will assume I work 60 days and fly 200 hours a year. That means I have to charge $60,000 a day and $17,000 a flight hour to cover my costs.

This is a reasonable rate of return and one that will, likely, happen many years. But if it's a bad fire season and I fly more than that, my company starts making a lot of money. The government doesn't like this. They think it's war profiteering and thus drill down to ensure our costs do not exceed a margin they find acceptable (which is basically 0 percent). But they give no credit to the fact that we invest hundreds of millions into our businesses, and they give no guarantee of use. If I don't fly at all in a season (which can happen), then I bear 100 percent of the costs and the government has no obligation to provide compensation. They think that's a fair deal.

Now, let's go back to the basic math on all of this. If I fly 120 days and 400 hours (which is a reasonable season and one that will provide appropriate initial attack coverage to our nation), at those rates, I make, not surprisingly, twice as much. But if I know that's how much I will be used, I can cut those rates by up to 35 percent and give the government a much better deal. They are paying almost half as much per gallon dropped as they would on the "Call When Needed" rates. It's no different than buying in bulk at Costco items that never go bad, like toilet paper (I can hear the groans from industry as I compare aerial firefighting aircraft to toilet paper at Costco, but bear with me). If the item has a reasonable shelf life, buying it in bulk can save you significant cash. Oftentimes up to 50 percent. Why? Because the vendor has a degree of certainty you are buying a lot, so they can stock it more efficiently, spend less time packaging and storing it, and with the knowledge they will sell four to five times as much, you, the customer, get a better deal. Running a big-box store is mainly a fixed-cost game. You have a building, utilities, and a staff that has to be there regardless of traffic. So, the cost line is relatively static. What is variable is the revenue line. Thus, the vendor is incentivized to give the customer a good deal in order to keep maintain a revenue stream. As we all saw during COVID, toilet paper is an American staple. It will not spoil and it *will* get used. You may as well buy a bunch of it and stick it in the closet, because eventually, you'll use it all.

The U.S. Forest Service prefers to buy toilet paper one roll at a time—and they don't mind paying up for it. And, to extend the analogy further, oftentimes, they sit down on the toilet and when they realize they are out of toilet paper, they do the embarrassed text from the bathroom, asking for an extra roll. Unfortunately, and perhaps inevitably, that that roll ends up costing a lot more than it should.

But change comes slowly, if at all. The metaphor of the ocean liner comes to mind. You can't turn it around quickly, which unfortunately leads some people (mainly those who benefit from staying the course) to believe that it's not worth changing direction at all. Too much effort is required. Too much time and money and energy. But the truth is, you can turn an ocean liner around—if you're truly committed to doing it. Change is difficult, sometimes even painful. Almost anything worthwhile comes at a cost. But in the case of wildfire management, the cost of doing nothing is greater than the cost of change.

★ ★ ★

Another story from the summer of 2015. One of our Twin Commanders was flying air attack with a modified camera hole in the bottom. This was rudimentary surveillance equipment—a camera system built into the belly of the plane—and was basically included in the price of our service. We were trying to show our clients the value of real-time infrared surveillance, which allowed us to scout and map fires from the air. So far, it had been a big success. Most of our clients loved the Twin Commander and the way it had been outfitted; they were impressed with the potential of real-time infrared surveillance in the battle against wildfires. But there were detractors, as is always the case with new technology.

A little background in terminology and process will help with the story. Yes, "air attack" describes the plane controlling traffic in a fire zone. But "air attack" also refers more pointedly to the Air Tactical Group Supervisor (ATGS), the agency official who is directing the show. He sits in the front passenger seat of the cockpit next to the pilot. It is the pilot's job to safely guide the ATGS through the fire zone so that he can focus all of his attention on the task at hand. There is no comingling of responsibilities—or shouldn't be, anyway. The air attack pilot does not give orders to the other pilot nor advice to the agency official he is transporting. And the agency official does not tell the pilot how to do his job, although he may request a flight path that grants better visibility.

Additionally, in the case of Bridger Aerospace, a third person sat in the back and operated the sensors and camera system. This person's flight time was also included with our fee. Basically, we were giving away our time and technology as a way of marketing a product—"Hey, look what we can do!"—in the hope that this would lead to both better fire management and expanded business for us. In the summer of 2015, those guys in the back seat were one of three people: me, another former SEAL teammate of mine named Kevin Brown, and, most importantly, our chief intern, Mike Tragiai. I met Mike in the fall of 2014 when I went to the gym one day at the Ridge Athletic Club in Bozeman. As I was checking in at the front desk, I gave my phone number. The young man behind the desk who was checking me in and folding gym towels took my number down.

"Minnesota, huh?"

"Yeah, Shoreview."

"I'm from Shoreview!"

It turned out that although we were several years apart in age, we had grown up about a mile from each other, and our families knew each other. He was a student at Montana State University, and he became the first intern of many that we would harvest from the school as per our founding thesis. Mike's interview consisted of him showing up on time and actually being put to work upon walking through the door. To this day, he is one of our top performers. Oh, and he also didn't get paid anything as an intern. So, with a twenty-year-old college student in the back, we were off to the races.

ATGSs are highly trained and experienced firefighters, having spent a minimum of ten years as a ground firefighter. Some are former smokejumpers. Rarely are they also pilots, but that's okay. They understand what a fire looks like on the ground and how it is most likely to behave. Our job as aerial firefighters is to support the ground effort, and the ATGS is the person most qualified to serve as a liaison in that effort. He has the experience of a ground fighter and the perspective (literally) of an aerial firefighter. He can see what is happening and relay that information to both pilots and ground crews.

"Watch out for the cliff off to your left."

"When you come up out of that drop, make a right-hand exit to the terrain on your left."

"We need a helicopter to support Division X-ray."

And so on…

Sometimes, in the case of a big or rapidly expanding fire, a second ATGS will sit in the back of the plane. Both officials will be on their radios simultaneously, giving directions to different crews. Another instance in which there might be two agency officials is when a younger ATGS is being trained by a more experienced agency official. Regardless of the circumstances, the ATGS is in charge of the FTA. And the Twin Commander that carries him is essentially a flying command post. That is the way it has always been.

In this case, our first plane, tail number 578WY—the Twin Commander purchased from NOAA, modified with our surveillance equipment—was flying on a U.S. Forest Service contract out of Region 1—Montana and Idaho. The ATGS assigned to the plane was a Native American named Ken. He was a big guy, well over two hundred fifty pounds, and a former smokejumper who had been fighting wildfires for pretty much his entire adult life. But he was open to new ideas and technology and embraced the infrared surveillance equipment on board, using it to call in helicopter drops on hot spots that were not visible to the eye because of smoke.

"That was pretty cool," Ken said after his first day using the infrared cameras. Chief Ken, as we all called him (because he was a big, commanding Indian), was a man of few words. Those four words were high praise coming from him. As I got to know Ken better over the years, I came to respect him more and more. Unfortunately, Ken eventually rotated out and was replaced by a different Fire Service official with a decidedly less enthusiastic attitude toward new technology.

"No, thanks," he said as we introduced him to the surveillance equipment before the mission. "Don't need it."

"Well, can we at least show you how it works?" I said. "You're already paying for it—I mean, it's included with the plane."

He shook his head. "Not interested. I've been doing this for years. I'll be fine."

We went off on the mission, and as we flew over the fire, he pulled out a crappy little smartphone and began taking pictures through the window, zooming in on the fire, cursing about the smoke and the lack of visibility and other predictable issues as he tried to text images to his guys on the ground.

"Ah, fuck! I don't have any service," he shouted at one point. "I'll have to wait."

I was in the back of the plane, operating the sensors for training purposes (and on the off chance that he might change his mind). His frustration, I figured, represented an opening.

"You know, we have this amazing system built into the plane that can do this for you. Look, it's got map overlay and all this other cool stuff. Why don't you give it a try?"

Without glancing back, he snorted disapprovingly.

"I don't need that crap. I'll just do it this way. It's what we've always done."

And then he went back to taking grainy pictures on his cell phone, holding the camera high and then low, desperately seeking a signal strong enough to transmit the photos. It was astonishing to me, the reflexive refusal of doing things differently, even when a different approach was so clearly warranted. This guy was sitting in a $150,000 plane outfitted with another $100,000 in high-tech equipment, all of which would have made his job easier and provided useful information for the other pilots and ground crews assigned to the fire. He was married to 2004 technology when 2015 technology was right there, wooing him, begging him for an opportunity.

"It's what we've always done."

That wasn't an explanation or even a counterargument. It was sheer laziness. And it was dangerous. Why wouldn't you use the best available technology if it can save lives and property? Part of the answer, I realized, was fear: *"If this technology is so good, then maybe I'd rather not use it, because it'll make me obsolete."* I saw this in the military as well.

Or maybe it was just good old-fashioned narrow-mindedness.

"I don't like new stuff. I like the stuff I already know how to use. Leave me alone and let me do what I do. I don't want contractors trying to sell me this new crap. This is a sales pitch. I'm here to do my gig. Don't pawn your junk off on me."

I have no idea what provoked his response, but it was a vivid example of the challenges we faced as a business offering innovative technology and of the rigidity of the aerial firefighting world. For all the heroic efforts of the men and women on the front lines—the pilots, ground crews, smokejumpers, mechanics, and other support staff—it remained a sprawling bureaucracy that would have to be dragged kicking and screaming into the twenty-first century. From radios to software to aircraft to safety gear, the desire for improvement could be seen in so many people, but it was constantly stifled by the disaggregated funding lines, authority, and alphabet soup of agencies with their fingers in the pot. It was, in many ways, reminiscent of the Department of Defense.

It reminded me of a confounding story about the Civil War. The Union ended up winning the war, not necessarily due to prowess, but due to the industrial capacity of the North to overwhelm the Confederate army. But even with the Union's principal advantage being manufacturing and technology, the primary weapon for the U.S. Army was muzzle-loading rifles. The scenes we have all seen in dozens of Civil War movies of soldiers standing in line jamming a ball and powder down their rifle while being gunned down and split in half by cannon balls are accurate. This brutal collision of metal and flesh took hundreds of thousands of lives. But the war could have ended much sooner. In 1860, the Henry repeating rifle was made available to the U.S. Army (the Confederates desperately wanted it but couldn't access the factories in the North). The U.S. Army declined to purchase the nine-round repeating rifle because it wasn't considered proven technology yet, despite frontiersman and cavalrymen using them with great success. Conservatively, the Henry repeating rifle would give a soldier the ability to fire over one hundred rounds to every single round from a muzzle loader. The generals knowingly prohibited better technology from getting into the hands of their young soldiers because they felt it was not yet ready for use. The result was tens of thousands of dead young men who were shot at point blank range while jamming a rod down their barrel instead of engaging the enemy. And untold thousands more died as this decisive technological advantage would have most certainly hastened the end to hostilities.

Unfortunately, this sort of response is not uncommon in the military. When I was going through Ranger training in 2005, we had one instructor who refused to put his night vision goggles on during training. He was an older fellow and I don't believe he had yet deployed during the global war on terrorism. We were traipsing through the Georgia woods at 2 a.m., tripping under our heavy loads and sleep-deprived status. It was the instructors' jobs to guide us along and ensure our safety, but this particular instructor seemed more like a trainee in his reluctance to properly utilize the technology at his disposal.

"Sergeant, this will be a lot easier for you if you put your damn NODS down," one of the younger Ranger instructors said to him, as if he were a student.

The sergeant scoffed. "Nah, I don't trust those damn things."

Even in the blackness of the Georgia wilderness, I could feel the eyerolls from the younger guys, probably half his age, who had already completed several deployments with Ranger battalions and understood the critical advantage of night vision goggles during battle. But it's tough to teach an old dog new tricks.

A similar attitude frequently permeates the U.S. aerial firefighting apparatus. All too often there is a reluctance to innovate and a refusal to listen to voices that may have insight into what's better for both citizens and firefighters on the ground. Some of this resistance comes from distrust of industry, and some of it is just bureaucratic inertia. But neither of those things is an excuse for not supporting those who need us.

The Yarnell disaster in 2013 was a perfect example of what enhanced sensor technology and information can do to help protect firefighters on the ground.

On June 28, near the town of Yarnell, in mountainous terrain outside of Prescott Arizona, dry lightning ignited a wildfire in desert scrub. The Yarnell Fire, as it was named, expanded to over 8,300 acres in three days, moving quickly in conditions marked by low humidity and high wind. High desert fires in the Southwest are characteristically erratic—they can travel fast and quickly change direction with a shift in wind or terrain. Although the fuel is lighter than that of timber fires, which can make them easier to suppress, the speed at which they move makes them remarkably dangerous. As part of the initial attack resourcing, the Granite Mountain Hotshots were dispatched as one of the many ground elements sent in to contain the blaze. The twenty-man crew was an Arizona–based team and was able to make it to the scene the next day.

★ MUDSLINGERS ★

Hotshots are among the most highly trained and capable first responders in the nation and are used for the most aggressive attack movements on a fire. This team was a Type 1 team, meaning the most qualified for high-risk wildfire tactics. The Granite Mountain team was directed to start making a line to protect the community of Yarnell, which was in the path of the approaching fire. As they were setting line, the lookout reported to the team that the fire was rapidly changing direction and would soon threaten their position. Communication issues inside the team and between the team and the adjacent units on the fire led to confusion as to exactly what units were in danger of a burnover, and what units were to continue building a line.

Eventually the team leader, Eric Marsh, realized the mortal danger his team was in with the approach of the new fire flank and directed the team to escape through a drainage. As they attempted to move, it was reported that several radio reports for assistance and information were either garbled, stepped on, or never received. At approximately 4:42 p.m. on June 30, the hotshot crew was in such danger that they laid down in place and deployed their fire shelters (aluminum foil metallic sheets that a wildland firefighter carries in case of a burnover). These shelters are meant to protect a fire fighter in the prone position from fast-approaching and inescapable wildfire—the firefighter's equivalent to a parachute for a fighter pilot, it is a weapon of last resort.

The team was not heard from after they deployed shelters. Two hours later, a local police rescue helicopter located the remains of all nineteen of the Granite Mountain hotshots. None survived.

This was an utter and perhaps unavoidable tragedy; however, it is illustrative of the impact of real-time information and how any ground tactical team can benefit from technology and communication. From Ed Pulaski hiding in a mine shaft in 1910, to the Mann Gulch fire in 1947, to Yarnell in 2013, fire can be a brutal teacher. We should be taking every advantage we can to stay ahead. Had the Granite Mountain Hotshots been better informed about the nature and growth of the Yarnell Fire, it's possible they could have escaped. Fire crews on the ground can benefit greatly from imagery, real-time maps, data analytics, and location data of other fire crews and ground assets. The technology is readily available and easily fits within current budgets, but the reluctance to act is structural. Sadly, a bureaucratic sclerosis infects the fire management agencies that is out of step with the pace of the threat they face.

★ ★ ★

Eventually, though, technology triumphs. Let me rephrase that: *good* technology triumphs. Our infrared camera system was the basis for a division of Bridger that proved too useful to be ignored for very long. There were pockets of acceptance within the Forest Service and other agencies, especially within the business community that contracted with those agencies. We had a good product. We just had to figure out a way to sell it. As I said, we started out by simply packaging the tech with our Twin Commanders. By the end of the summer of 2015, however, the surveillance portion of Bridger had spun off into a separate corporate entity: Ascent Vision Technologies. So now we had two businesses with a grand total of eight employees. I was still trying to fund both companies mostly with credit cards, scrambling from one day to the next, just trying to keep my head above water. And while it was exhausting, it was also thrilling, as there was a definite sense that we had begun to gain momentum.

The decision to sell our technology under a different banner was primarily based on two factors: branding and (as usual) economics. Invariably, some of the customers we wanted to address were going to be competitors with Bridger—specifically, other companies that flew airplanes, and probably even aerial firefighting companies. In the same way that General Motors does not want to buy parts or technology from Ford, our competitors were not going to be interested in buying surveillance systems from Bridger Aerospace. But if we took that same technology and sold it under the umbrella of a different company, then we'd not only be able to use it in Bridger products, we'd also be able to sell it to Bridger's competitors. You might reasonably ask if this isn't just a matter of smoke and mirrors—is the aerial firefighting world so big that no one would notice the connection between the two companies (like the fact that I was CEO of both)? And the answer is, of course some of them knew—maybe most of them. But optics matter.

Second, and equally important, was the issue of funding. We continued to burn through money while trying to grow both the firefighting and technology ends of the business. At that time, we realized there was greater financial potential in the tech sector. Yet gaining additional funding for it proved almost impossible. Banks like assets, and now Bridger had assets. We had planes and other equipment, and we had government contracts, which are among the most reliable and sturdy revenue streams. So we could borrow against those assets. Banks, however, do not like lending money to support technology. Conversely, private investors and venture capitalists love technology and have no interest in the airplane industry with its slow growth and low margin. Silicon Valley is run by investors taking big swings on companies that have almost nothing tangible to leverage. It's all about *potential*.

We had two viable products with vastly different profiles and needs, so they became separate but connected, which turned out to be the best decision we could possibly have made. Each was now free to pursue its financing needs through the channels that made the most sense. We were able to secure additional financing for Bridger through American Bank, whose CEO, Bruce Allen Erickson of Montana, had been a big aviation buff before he died in a helicopter crash in the fall of 2015. The influx of cash allowed us to expand our fleet of air attack planes, secure more government contracts, and continue to build a business that was rapidly becoming profitable. But most of that profit went right back into the business—specifically, into AVT. I didn't take a salary for the first couple years; Sam Beck took no salary the first year and a meager $1,000 a month in the second year. We stayed lean and mean, hoping that eventually, everything would work out. Growing the business was half the fun.

★ TIM SHEEHY ★

I reached out to a former teacher of mine from the Naval Academy, a guy who was five or six years older than me and a former Army Ranger, West Point grad, and standout lacrosse player. He'd been in the medical sales industry for several years, climbing the ladder and making good money, but he was getting bored and looking for a change. He understood technical sales and business growth. He was personable—the kind of guy you wanted to hang out with—but also a tireless worker. I figured he was the right guy to take AVT to the next level. And he was, although I never imagined that level would turn out to be as high as it was or that we'd get there so quickly.

In 2020, AVT, a company started just five years earlier with only a handful of employees, was sold to a U.S. defense contractor for $350 million. Along the way, our technology produced the first drone ever to be used on an active wildfire. Drones, procured from the military, had occasionally been used "off-the-books," but this was the first time a drone had been assigned to a fire zone. In general, drones are considered a nuisance at best and dangerous at worst in a fire zone, as they can interfere with the delicate choreography of multiple aircraft. But drones are like any other technology: utilized prudently and with expertise, they can be highly effective. Most importantly, AVT was critical in the design of what would become the DoD's most effective counter drone system ("drone killer," for easier understanding). Once again, a well-timed and abrupt pivot from airborne tech to ground protection capability proved to be the inflection point for one of our young businesses.

It was a hard decision to sell AVT, but I realized I had to focus on what my priorities were going to be for the next ten years. I had a lot of employees depending on my decisions, and most importantly, I had four kids and wife that needed some more of my attention. In the early days of AVT, I absolutely loved supporting the warfighter. Sitting with the Marines in the Yuma desert, trying to help them create systems to protect themselves better, saving lives—just like Bridger. But as our contracts grew and our business matured, I realized being an executive in the defense industry was not my strength. I was getting further from the warfighter and becoming a program manager. The team and my customers needed a larger company to steward what we were becoming. As Bridger continued to grow, the mission remained potent. It was clear that was where my efforts were needed most.

The proceeds from the sale were an incredible stimulus to the staff at AVT (all of whom had equity) and many of the staff at Bridger, who also were part of the early growth of AVT before it spun off. It also gave me the financial liquidity to support more initiatives at Bridger. The growth at Bridger was less explosive, but steady. And the mission continued: to become an integral and innovative part of the aerial firefighting community.

CHAPTER 9

FLYBOYS

To the layperson, it might appear that flying is flying—that the skills acquired and developed in one type of aeronautical endeavor would be transferrable to another, and they are to a certain extent. But it would be wrong to assume that any experienced pilot would make a good aerial firefighter. Not everyone can make it through the U.S. Navy's TOPGUN program and become an F-14 fighter pilot, after all. So why would someone assume that everyone who has ever flown a plane has the ability to be an aerial firefighter? It is, in fact, a uniquely challenging and dangerous type of work. Not everyone is cut out for it, glossy resume notwithstanding.

When Bridger first began looking for pilots in the spring of 2015, a fairly significant pilot shortage had impacted virtually all segments of the aeronautical industry. We had found a couple pilots through my military connections, and they in turn reached out to their contacts in the hopes of allowing us to have between four and six pilots available in the first season. Tim Cherwin had flown commercially for some regional airlines for a while after leaving the military and recruited one of his friends, a guy named Richard. Richard was great on paper and extremely competent in the cockpit of a commercial jet, a captain who had amassed some fifteen thousand hours of incident-free flight time. Now he was in his early fifties, somewhat bored, and looking for a new challenge. Fighting fires, he thought, sounded like a great way to expand his skill set and do something different and meaningful with his career while he was still young enough to make a change. His intentions, I believe, were 100 percent honorable. He should have been a strong candidate to join our team.

But as it turned out, Richard had no idea what he was getting into.

He flew out to Montana as we were preparing for the first fire season. The training was not particularly intense or extreme. We hadn't acquired any scoopers yet, just the two Commanders, and we were simply putting them through their paces, flying through the mountains in relatively clear weather and only skirting the edges of fire zones since we were merely training and not on assignment. Nevertheless, it was a completely different type of flying than Richard had grown accustomed to—scooting in and out of canyons, quickly changing altitude, dealing with weird mountain turbulence and reduced visibility due to nearby smoke and flames. In short, training that was intended to provide a taste of the sort of conditions we'd encounter while flying air attack on a fire.

After two days of flying, just as I was starting to think that everyone was settling in nicely, Richard walked into our little office and quit.

"Sorry," he said. "But this low-altitude stuff, in and out of the mountains? With all the smoke and wind? It's not my cup of tea. I'm out."

Sometimes you know better than to try to talk someone out of a decision. I could see by the look on Richard's face that his mind was made up. He looked, for lack of a better word, beaten.

"You don't have to pay me for my time here," he said. "I know I'm putting you in a tough spot, and I feel bad about that. But this job is not for me. I am not comfortable with this operational environment, and I don't think I'd be an asset out there."

I did not disagree. Neither did Tim. The last thing you want when you are fighting a fire (or starting a new aerial firefighting business) is a pilot who not only doesn't want the job but doesn't think he's up to meeting the demands of the job. The interesting thing was, I hadn't even noticed that Richard was struggling with self-doubt. He had done a pretty good job of hiding it until that very moment.

"Okay," I said. "Thanks for being transparent."

He shrugged. "Yeah, I guess." Then he walked out the door, got in his car, and drove away.

At that point, I realized two things: 1) the industry-wide pilot shortage was going to be more problematic than I had realized, and 2) perhaps we were better off recruiting pilots through the more traditional military and bush channels, both of which demanded skill sets that were more easily transferrable to the rigors of aerial firefighting. It's understandable that a commercial pilot might seek the excitement of aerial firefighting, and it's admirable that he or she might want to pursue a type of flying that is more obviously service-oriented. But the fact is, not everyone is cut out to be a first responder. There is tremendous turnover in all segments of the wildland firefighting community. It is hard, dangerous work that does not pay all that well. There has been a more recent push to fund wildland firefighting jobs in a manner similar to positions with municipal crews in structural firefighting. Better compensation, benefits, and job security would certainly help reduce turnover. But the fact remains that finding competent pilots for a job that is perhaps the most dangerous in all of aviation—in some ways equal to combat flying—will never be easy. There is simply a limited number of people willing and able to do the job.

The point of this anecdote is not to cast any sort of aspersions on commercial pilots. Most of them are highly capable aviators, and a fair number have military backgrounds. But here's the thing: while airline training is thorough, it is intensely focused on risk mitigation, procedures, and is guided by a zero-defect mentality. Commercial aviation training goes to great lengths to avoid any situation that could enhance risk that hasn't been modeled and practiced repeatedly—anything that isn't in the training manual. Pilots will only fly into airports that have a certain kind of approach, a specific type of runway; they are always in radar coverage and abiding by a specific flight plan that is not to be altered except in an emergency; they are constantly under the watchful eye of various air traffic controllers. Nothing is left to chance. It's as close to a controlled environment as you can possibly have in aviation.

And it is for that reason that commercial air travel is the safest form of travel. Nothing else comes close. When something does go wrong, it is altogether shocking, not least because it is such a rare occurrence.

For example, in 2013, Malaysia Air Flight 370 disappeared after departing Kuala Lumpur, bound for Beijing. The aircraft never arrived at its destination and a massive search ensued, beginning with the region immediately surrounding the flight path and eventually expanding to cover the entire world. To this day, it is still unknown what happened to Flight 370 and its 239 passengers and crew. The incident captivated the world for weeks and months before fizzling out without an answer. Most people were left aghast that this could happen. A massive airplane full of hundreds of people can just disappear into the ether in the twenty-first century? The truth is, catastrophic accidents were all too common during the first seventy years of aviation history. Travelling by plane was risky, no two ways about it.

But what is even more astounding than the lack of early safety is the speed at which the industry travelled from being highly risky to being the safest and most efficient form of travel ever known. It is truly amazing how safe commercial air travel is today when you consider the scale, complexity, distances, and speeds.

Commercial airline pilots are skilled technicians whose job is to guide the aircraft through a mechanized process and to make sure that everything goes flawlessly. If something goes wrong, their training will allow them to step in and take over—but again, that rarely happens. Commercial pilots are like surgeons at a major metropolitan hospital. Everything is scripted and controlled; everything is sterilized. They're supported by a highly trained and professional staff. The entire process is intended to be predictable and boring. That's the goal. Because boring is safe. Boring means nothing went wrong. Commercial pilots are trained to precision and asked to manage flights that are conducted with extreme predictability. On the rare occasion that something goes wrong, their job is to get the plane on the ground as quickly and safely as possible. It's a very specific type of training and the results speak for themselves. Airlines indisputably provide the safest form of travel.

Aerial firefighting is anything but predictable. While there is a certain familiar choreography to the fire zone and a carefully orchestrated plan of attack, the nature of the job is such that…well, shit happens, and you must adapt. Everything about training to fight fires is predicated on the fact that the unexpected should be expected.

For that reason, above all others, airline pilots generally don't do well when they try to make the switch to aerial firefighting (there are exceptions, of course, and we've had a few of them at Bridger). The work environment is the complete antithesis of the orderly world they are accustomed to. Aerial firefighters are basically operating in an emergency scenario the entire time they're in the air, flying through mountains that are literally on fire. As noted previously, in commercial flight, if you fly within a mile of another aircraft, it's considered an egregious error. Alarms go off, everyone immediately banks away from each other, and there's a big investigation afterward. When you're in a fire, it's not unusual to be within a couple hundred meters of another aircraft. You're covered in smoke and battling fierce, unpredictable winds at low altitude. At all times, you are close to other planes, close to the smoke, and close to the ground, which also happens to be on fire. You're always flying visually, rather than with instruments. The entire environment, if you were in an airliner, would be considered an emergency. Yet it is merely a day at the office for aerial firefighters.

Airline training is excellent; it does what it's supposed to do. It minimizes variables and drills procedure in a very structured and well-resourced environment (air traffic control, dispatch, control, maintenance, flight crew). But our mission is different, and our pilots have to be constantly responsive to a chaotic environment. We can't really minimize variables because our entire mission is a variable. Your mindset shifts from specific mission planning to a more skills-based mission preparation. It's actually very similar to how SOF evolved in Iraq and Afghanistan. In the previous era of SOF, a team would literally go into isolation and specifically plan each aspect of every mission for weeks or months at a time. The op tempo was slow and deliberate. A lot of SOF teams were available for just a few operations, so preparation space was high. The OIF/OEF construct was a complete reversal. Now, demand outstripped supply. Instead of days and weeks to plan a mission, you had minutes or hours. So specific mission planning was no longer possible. You didn't pack and configure your gear for each operation; you had a standard load out and you used standard operation procedures and training to manage each mission as it happened. It was a different construct that heavily relied on a decentralized responsibility structure and a highly trained talent base. This is very similar to aerial firefighting. You can't load the approach and brief the pilots and manage all the risk of a fire when you don't even know where it is until you're flying toward it.

In an airliner, the seat belt sign comes on five minutes before you hit turbulence because you know it's coming. There are certain types of radar, as well as communication from other aircraft, that detect rough air well before it hits the plane. There is time to prepare and perhaps even adjust altitude to avoid it. Aerial firefighters have almost no clue what they will encounter on a given mission other than that there will be smoke and wind and fire and frequently bone-jarring turbulence. You make one snap decision after another based on the information you have in front of you and the time available. Sometimes a fire zone can be relatively calm. By the time you arrive, ground crews are already on the scene, the weather is cooperating, and soon the aerial crews are in a rhythm. Air attack is neatly coordinating a dance between the helicopters, scoopers, and tankers, and everyone must do their job flawlessly. Those are the good days, when a fire becomes almost predictable.

Almost.

But you're just as likely to arrive and discover that the fire has jumped a line and is spreading like crazy. The wind is kicking up, and the smoke is making each drop progressively more hazardous. It's just the nature of the job, and you either adapt to it or find another line of work. I know a lot of airline pilots who aren't crazy about their jobs.

But they like the security and the pay and the benefits. Like Richard, they might be looking for a different challenge, but when faced with the reality of flying into a fire—for a lot less money and a lot less security—well, it just doesn't seem worth it.

We generally recruit our pilots from the military and the bush community—people who are used to flying under duress in parts of the world that depend on aviation just for survival. These are men and women who are wired differently. Military vets, in particular, tend to still have the adrenaline flowing. They don't want a slow and predictable job; they want to be out there mixing it up. Many of them (and I include myself in this group) want to be part of a mission serving something beyond themselves. I know of more than a few veterans who made the jump from firefighting to commercial aviation. They left thinking they had reached nirvana. Most of them became bored and disillusioned and returned to firefighting.

"It gets in your blood," said Bob Forbes, who flew tankers for nearly five decades before retiring in 2016. "Talk to anyone in the industry—ground fighters, pilots…doesn't matter. Fighting fires is not just a job. It's a way of life."

Although no longer a full-time working pilot, Forbes helped out on a few fires as recently as 2021, as a copilot on a Chinook helicopter.

"I guess you'd say I'm retired, but not completely retired," Forbes noted. "And I'd work even more if I could."

At eighty-two, he knows that's unlikely and perhaps not advisable, but Forbes has had a long and distinguished career, one that spanned some of the most dangerous periods in aerial firefighting, during which he lost dozens of friends and fellow pilots. But he regrets nothing and would do it all over again. When asked whether it's the flying or the firefighting that makes the job so appealing, Forbes laughed.

"Well, it's both, I guess. If you're working around populated areas, it's like, 'Let's see if we can save this house. Let's see if we can save these people. Let's see what *good* we can do.' It's important work. There have certainly been times where I thought, 'Holy crap, I'm not coming back.' But I've been one of the lucky ones."

Like most pilots who have long careers as firefighters, Forbes was never a cowboy, but rather someone who accepted the risks that came with the job and tried to perform at the highest possible level while still returning home in the evening. That permeates the industry to this day.

"It's kind of like a regular job for me at this point," said Bridger's chief pilot, Barrett Farrell. "I mean, we know the risks are there. But they're calculated risks. I will not deliberately go down into a canyon that I don't think I can get back out of. We don't take stupid risks or stupid chances. It's as simple as that. We're going to do everything in our power to help manage or suppress the fire, but we all want to come home at night. So we're not going to take unnecessary risks. The majority of [firefighting] pilots are actually pretty laid back. I think you kind of have to be that way if you want to have a long career. If you're constantly worried about what we are doing and the risks and the danger it would drive you crazy. The risks are there—absolutely—but you learn to handle the stress of it."

In some ways, being an aerial firefighter is similar to being in special operations in that the public might perceive SEALs and Green Berets and Delta Force operators to be cowboys, but nothing could be further from the truth. Adrenaline junkies, perhaps. But not individualists. There is a profoundly deep team mentality in Special Operations, just as there is in the firefighting community. You really don't want the lone wolf kind of guy in either of those settings. You don't want the guy who thinks he's Jason Bourne next to you when you're trying to clear a building in Afghanistan or Iraq. You don't want that guy circling near you when you're in a fire zone. And you sure as hell don't want him in the pilot seat when you're the copilot. There are tactics and procedures that govern the fire zone—everyone must know their job or people get hurt and fires rage out of control. Sometimes that happens anyway, so why worsen the odds with rogue behavior? In the SEAL teams, if someone selfishly violated the rules and put teammates at risk, they were gone.

See you later, go play a video game or become a mercenary. You're not working for us.

It's the same with our business. We've had some of those guys show up for training or interviews, and it's like, "Listen, man, we're not trying to be risky here; the job is already risky enough as it is. Frankly, what we're trying to do is do what the airlines do: engineer risks to the minimum possible level. We don't need anyone going off on their own and flying lower than they need to or doing some fancy tricks. We need you to be part of the team. That's all."

It's a delicate balance. You want pilots who have experience flying in challenging environments like bush flying and combat aviation. And some of those guys, quite honestly, do have a bit of an edge. The bush pilots, in particular, sometimes grow accustomed to operating on their own, and for them, the individualist persona is hard to shed. But the aerial firefighting community is largely self-selecting: the lone wolf doesn't last. Neither does the cowboy. Optimally, they leave before they get hurt…or hurt someone else.

★ ★ ★

Aerial firefighting is much safer than it was in its infancy, when guys like Vance Nolta were flying virtually blind into fire zones and dropping water from makeshift tankers. Advances in technology, communications, meteorology, and mechanical integrity have combined to reduce the likelihood of a catastrophic incident. Nevertheless, they do still occur with some regularity. From 2000 to 2013, for example, nearly three hundred pilots were killed while fighting fires—an average of over twenty per year.

While there is no question that allowing planes to fly at anything less than optimal mechanical condition is an egregious dereliction of duty (and there is a history of this in our business, which we will discuss in subsequent chapters), it is also true that most crashes are the result of pilot error stemming from the fact that the job is conducted under extraordinarily challenging and dangerous conditions. Wildland firefighting pilots routinely fly at low altitude in a smoke–filled area with numerous other aircraft in the vicinity. They fly in mountainous terrain in changeable weather.

It is, quite simply, a job with a very low margin for error. Mountain flying is the most unique and difficult type of flying in the world primarily because the weather in the mountains is so unpredictable. It can be calm and perfect on one side of a mountain, and then two minutes later, you go to the other side and you're getting thrown around the cockpit by fifty mile an hour wind. As every pilot knows, mountains do strange things to the wind. You can get a downdraft of two thousand feet a minute that will literally drive a plane into the ground—it's like the hand of God coming down on top of you and slamming the plane into the earth.

And, of course, fire will be exacerbated by those conditions. Dry, windy weather in high mountainous areas just makes the fire go farther and faster and harder. The more intense the fire becomes, the more likely you are to experience a phenomenon in which the fire creates its own weather system. You can be flying in completely clear weather and then suddenly fly into a valley with a raging wind–driven fire, and suddenly it's like you're inside a washing machine, getting bounced all over the place as the hot air coming off the ground collides with cooler air above.

Couple that with the extreme changes in altitude faced by pilots in the Mountain West, and you have a flight environment that can whiten the knuckles of even the most experienced pilot. In an airplane, the higher you go, the lower the performance of the airframe because the air is thinner. So, if you're at sea level and you go to full power on an ascent, you're going to get twice the rate of climb than you do if you're at fifteen thousand feet. Temperatures have a similar effect on the air frame. That's why a plane can take off with more people, more fuel, and on a shorter runway at 2 a.m. on a cool evening than if it were flying at 4 p.m. when it's a hundred degrees. And, of course, in wildland firefighting, it's always hot and sunny. Throw in unpredictable winds—at times gusting to near tropical storm force—and you basically have the most hazardous aviation environment imaginable.

Accidents don't always happen under duress. Sometimes the very things that make someone an exceptional pilot—experience, spatial awareness, skill, confidence—can undermine his own safety.

"You can sort of equate it to swimming," said Barrett Farrell, who has been flying planes since he was a teenager. "At first, you're a little cautious. And then you get stronger, more comfortable, and maybe you take some chances, and then suddenly you're out on a river and you can feel the current pulling you away. And you realize, 'Hey, I have to be careful here. I'm getting too comfortable.' It's kind of the same thing with firefighting. You forget that it's dangerous, or at least you don't think about it as much."

Farrell paused and smiled.

"But then you get a reality check."

For Farrell, it happened in the summer of 2021 while the crew was stationed at a fire base in Chico, California, where he was copilot on a Super Scooper with Captain James Stewart. The pair were twenty drops into a fire one afternoon, everything going smoothly, when the reality check hit.

"We came in for the scoop, into the wind, the way we'd been doing it all day," Farrell recalled. "And then the wind shifted and kind of worked around us a little bit, pushing against the tail, just as we were about to scoop. The wind got under one of the wings and lifted us, got us off balance, and we started hitting the water. And all of a sudden, I heard Jim yelling, 'Pull up! Pull up! Pull up!' And we immediately aborted the scoop. Fortunately, Jim had full thrust control, so we were fine. We went back and up and figured out a new spot where we could scoop with a different approach based on the direction of the wind. We told the other pilots what had happened and continued making drops. It didn't end the day, but it was definitely a little bit of a wake-up call. And it came out of nowhere."

Scoopers are seaplanes, of course, so they are capable of stopping in the water, but not at high speeds, not unexpectedly, and certainly not while listing to one side or the other due to wind shear.

"If we had kept going for even just a few more seconds, we might have been in too deep to pull up," Farrell recalled. "The wind could have spun us around, and…"

He did not complete the sentence, but the implication was clear. You never want to be too comfortable—and certainly not complacent—in the cockpit when you're on a fire because everything can change in a heartbeat.

"Once in a while you get a little reminder that, oh yeah, this is a dangerous job," Farrell said. "But you do get comfortable being in the smoke, being around the fire. I always try to remind myself not to push it because I want to go home to my wife and kids. That's the most important thing."

But pilots are pilots and firefighters fight fires. Even the most experienced professional among us will sometimes find himself wrestling with common sense and restraint. Here, occasionally, the air attack officer can serve as both traffic coordinator and arbiter. From his vantage point high above the fire zone, the air attack officer is not necessarily aware of the turbulent conditions faced by tankers and scoopers as they swoop into a canyon. But experience and a finely tuned ear can help save a pilot from his own worst tendencies.

"Because they're so high up, air attack doesn't necessarily know how bad the conditions might be," Farrell said. "They can't really tell if I just pulled three Gs on a turn and maybe I was a little too aggressive. What they can do is listen carefully to the sound of the pilots' voices. They can read between the lines. For example, a pilot is not likely to say, 'I don't want to go in there right now; I'm pulling out.' Honestly, nobody wants to be that guy. But you might hear a pilot say something like, 'Hey, it's getting pretty rough in here, be careful.' And if air attack hears a few people say that, then he'll know it's a cue that the pilots are getting uncomfortable with what they're doing, and he has the authority to make changes so that everyone is safer."

Of all the factors that can conspire to take a pilot's life, perhaps none is more of a threat than hubris. This was true half a century ago, and it remains true to this day. Aerial firefighters enjoy (or are burdened with) a greater degree of autonomy than most other types of pilots (certainly far more than commercial pilots), and their desire to do legitimately good and important work as first responders, combined with a deeply ingrained sense of competition, can lead to problems. Jim Barnes is a U.S. Navy and Marine Corps veteran who served during and after the Vietnam war, eventually transitioning to firefighting when he was thirty years old. A native of Northern California, Barnes was flying cargo planes out of San Francisco when he was recruited by a friend from the aerial firefighting world, and he soon found himself flying air attack out of Rohnerville Airport in Fortuna, CA. He was thirty-one years old when he worked his first fire and spent the next thirty-five years flying tankers under the direction of CAL FIRE.

Barnes was among the agency's first S-2T pilots, and he later became an instructor as well. When asked about what the job was like in his early days, he responded by saying, "Well, I got used to delivering eulogies. Let's put it that way. For a while, we were losing an S-2 pilot about every other year. I think I personally knew about forty pilots who were killed."

Barnes was in the cockpit long enough to not only witness disasters, but also to experience the myriad changes that have made the pilot's job safer. Yet he was quick to note that the job will never be without risks, and those risks are sometimes exacerbated, paradoxically, by a pilot's confidence and skill.

"Things happen unexpectedly," he said. "They come on you fast, and then they're over with. Hopefully you handle it right. It's usually a perception problem while you're trying to figure out how you're going to drop on a target. You're in a tight canyon and there's smoke and fire and everything else. You're focusing too much on what you're trying to do, and then you lose your situational awareness. The next thing you do is look up and see a wall of trees in your windshield. Those are the kinds of things that can kill you."

When asked if a pilot's natural inclination is to err on the side of caution, Barnes chuckled.

"Nobody thinks they're going to die. It's always the other guy, right? But if you fly tankers long enough, you eventually realize, 'Yeah, it could be me.' I've walked through the ashes of tanker pilots who were far greater than me, so I was acutely aware of what could happen. I knew it could happen to me. And I think if you don't believe that, then you're in danger."

Barnes also acknowledged the issue of temperament—specifically, the pilot's natural competitive instinct.

"If you're trying to compete with another tanker, it's a problem," he said. "We had our sharpshooters, you know? Guys who walked on water—or thought they walked on water. They would try to thread the needle every time they went out, and if you try to compete with somebody like that, you could set yourself up for a fall. But the truth is, there is absolutely competition among pilots. It's not something that is discussed openly or directly, but everybody knows who the best guys are. My ambition was just to hang in there and do a good job. I'd look around and think, all these guys are great. I just want to be one of them. I'm not going to try to outdo them."

Barnes' career was long enough that he saw numerous positive changes in the industry, especially regarding pilot safety. Wildland firefighting pilots will always have a bit of an edge—it comes with the job—but today, they are more team-oriented and supported by immeasurably better equipment and a healthier work environment.

"There used to be a lot more cowboys than there are now," he noted. "I think it was the culture at the time that promoted it. The people there now are not aware of what we had when I started out. Honestly, it was a toxic culture, and this was true throughout aerial firefighting. We were running a junkyard air force on a shoestring budget. We still had guys from World War II flying tankers. We still had the B-17s. And, you know, there weren't a lot of regulations. There was a lot of what I would call rogue pilot behavior. We had guys that would take chances. Guys who had an attitude that was like, 'Hey, I've flown in the military, and this ain't shit compared to that.'"

The truth, of course, is that it's just a different kind of shit.

"We had some incredible pilots," Barnes said. "Guys who were flying S-2s in the Navy during Vietnam. These were carrier pilots. But some of them would come into aerial firefighting and have real problems, mostly because they'd fly too slow and have issues with stalling out. They were used to landing on ships at eighty-five knots, and then they'd go out and try to drop on a fire at roughly the same speed, and it just wasn't fast enough. It was negative transference—you know, thinking that something that works in one situation will automatically be applicable to another situation. But the training is also key in this issue. Back in the day, there wasn't a lot of training. I worked for one company where the training was: if you could fly the plane from home base up to the tanker base, you were good to go. That was your training."

Barnes paused and laughed.

"Unfortunately, I'm not kidding. But as I got into training and became an instructor, I saw changes. We brought in some great Navy pilots, Air Force pilots, guys who really knew how to conduct training and how to evaluate what went wrong and not do it again. That's how we found out we were flying too slow."

Old dogs can, in fact, learn new tricks, and even the most talented, stubborn, and resolute pilots, if they're smart, will accept the fact that the biblical proverb is true: "pride goeth before a fall." Among the most legendary military and wildland firefighting pilots of the last half century was a man named Joe "Hoser" Satrapa, who flew F-14 Tomcats and F-8 Crusaders and completed a remarkable 162 combat missions during the Vietnam war and made an estimated 500 arrested landings on board aircraft carriers during his illustrious military career. A 1964 graduate of the U.S. Naval Academy, Satrapa was involved in the development of

★ MUDSLINGERS ★

the Navy Fighter Weapons School, also known as TOPGUN, and his legacy bears comparison to some of the larger-than-life figures in U.S. aviation, such as Chuck Yeager and John Glenn.

Intelligent, iconoclastic, gifted, courageous, and profane—all in roughly equal measure—Hoser earned his nickname one day while flying over the California desert during a training session in his F-8 Crusader. As his team of four jets approached the gunnery range target on their third pass, Hoser (who had failed to hit a single target on the three previous attempts) unexpectedly pulled out of the trail position, raced to the front, and unloaded all his remaining ammunition from an altitude of two thousand feet, about one and a half miles from the target—"hosing" off all his bullets in a single pass. Satrapa was instructed to return to the air base at El Centro, where his admonishment for the stunt was simply an unofficial declaration from his flight leader that he would henceforth be known as "Hoser."

A big man with an even bigger personality, Hoser was the kind of guy who made the Great Santini seem restrained. His lack of interest in both protocol and paperwork, along with his irreverent attitude toward non-aviators (including his superiors), contributed to Hoser being pushed out of the Navy in the early 1980s. He briefly went to work for CAL FIRE as an air attack pilot before being summoned back to military service by then-Secretary of the Navy John Lehman, a former crew member of Satrapa's in the Navy. Improbably, Hoser was given a retroactive promotion to commander and assigned to the new Navy Fighter Weapons School in Miramar, California.

Anecdotes about Hoser's outsized personality and exploits are legion, but perhaps none is more illustrative than the one involving an accident that could have permanently grounded him. Hoser was overtly fond of weapons and carried a veritable arsenal of handguns, grenades, and knives with him whenever he flew on a mission in the event that he would have to defend himself if he crashed or ejected in enemy territory. He was, reportedly, tinkering with adapting a 20 mm cannon for use in an F-14 Tomcat when an explosion ripped off his right thumb and forefinger. The index finger was no big deal, but the loss of his thumb was devastating, as a pilot uses their thumb to control a jet's trim tab. Undeterred, Hoser convinced a surgeon to amputate one of his big toes and graft it to his hand as a makeshift thumb. While ungainly in appearance (it left Hoser with three fingers and an oversized thumb—sort of like a mitten), the hand worked well enough that Hoser was able to continue flying. And he did so for many more years, adding a stellar career as an aerial firefighter to his already impressive military resume.

With CAL FIRE, Hoser became known as Joe "in the smoke till you choke" Satrapa, reflective of the same do-or-die attitude that had carried him through hundreds of missions in Vietnam. Among the often-recounted stories (which Hoser himself loved to tell) is the time that his S-2 Tracker was doused in retardant by a tanker flying above him. So thick was the mud on his windshield that Hoser lost visibility and was compelled to open a window and reach outside to wipe the glass clean with his hands.

Hoser made it back safely that day, and every other day, despite sometimes pushing the limits of safety and reason.

"Hoser was probably one of the finest jet pilots ever," noted Barnes, who flew with Satrapa for CAL FIRE for several years. "But when he first got into flying tankers, he didn't try to tell everybody else how to do it. He sat back and watched. And eventually, he became one of the best instructor pilots we had. He was highly acclaimed for that.

"But in the end, he almost lost his life. He hit some rough air on a drop when he was flying real low, and he clipped a tree—cut right through his wing. Two more inches to the right, and the wing would have come off. But he was lucky, and he made it back. Of course, in aviation, it's just like firefighting—you make a joke out of it. We called it the Hoser Stall. But honestly, I think he was kind of born again that day."

For the less fortunate, there is no chance for a second life. Pilot error or mechanical failure are almost always the official cause of any accident, but there is a randomness to aerial firefighting that will leave even the most experienced pilots scratching their heads and thinking, *There but for the grace of God ...*

Barnes and Satrapa were both flying S-2 tankers out of Grass Valley, California, on August 27, 2001, when they were dispatched to work on the Star Fire roughly halfway between Grass Valley and Columbia. Two other tankers were also assigned to the fire, and as Barnes drew closer to the destination, he grew concerned about thickening clouds and smoke and diminishing visibility.

"The closer I got to the IP [initial point], the greater my apprehension became," Barnes would later write of the incident on his Facebook page. "A couple times I thought that I should abort the mission because of the bad visibility and return to base. I stayed in close communication with Hoser, who had come into much clearer air to the southwest. I was sure that sky conditions would improve as I neared the IP. They didn't."

It was only after a successful drop, when he was on his way back to Grass Valley, that Barnes allowed himself to think about the risk he had just taken:

"My thoughts brought me to a harsh reality. I had violated all the safety standards for visibility and for common sense. I had succumbed to 'mission creep,' even though there was no urgency to continue. I surmised that I must have flown through the heaviest area of obscuration because Hoser had reported better conditions to the southwest and the Columbia tankers did not report a problem. Despite the hair on the back of my neck bristling for what seemed like forever, I ignored all the warnings that should have been heeded. I came to the realization that by ignoring all the warnings, I could have been the cause of a mid-air collision at the IP."

Incredibly and tragically, when Barnes returned to the base at Grass Valley, he was met with the news that while he was working the Star Fire, there had been a catastrophic accident at a different fire in Mendocino County. Despite clear conditions, two S-2 tankers collided while dropping retardant on the Bus McGall Fire near Ukiah. Pilots Larry

Groff of Santa Rosa and Lars Stratte of Chico were both killed in the crash. Both were good friends of Barnes.

"We were all in shock, even though our lead instructor, Rich Ruggiero, had made the grim prediction in training that year that our next serious accident could well be a mid-air collision," Barnes wrote. "It seemed so ironic to me that all the contributing factors that could have led up to a fatal accident were operating at our fire—continuing into bad visibility, converging on a single point defined by GPS coordinates—by three other airtankers all at once.

"At the exact same time, one hundred miles to the west, airtankers operating in clear conditions had had a catastrophic midair [collision] resulting in two deaths. It didn't seem to make sense to me at the time, although we all knew about the potential danger. Tanker Captain Jim Cook had been working on a procedure for deconfliction in the circuit from the airport to the fire incident and back. It amounted to having protected airspace while both en route and at the fire. It incorporated required position reports, assigned altitudes, speed restrictions, and three separate clearances. Jim's procedures were memorialized in a document that became known as 'the Cookbook.' [At the time of the accident, it] was under consideration but had not yet been implemented."

A few days later, Barnes walked the crash site with a small group of people from the firefighting community, including Larry Groff's wife, Christine. On the road below, the crash site was a pile of rocks. Protruding from the pile was a handwritten note on a piece of paper. It was covered with hearts. The author was a young girl who lived nearby and whose family's house was among those that had been threatened by the Bus McGall Fire. Christine Groff picked it up, read it silently, then handed it to Barnes and asked him to read it aloud for the entire group to hear.

"I don't remember the exact words," Barnes recalled. "But it started with, 'Thank you. You have made the greatest sacrifice for all of us.'"

CHAPTER 10

THE SUPER SCOOPER

The first time I saw a Super Scooper was in the summer of 2015 in Lewiston, Idaho, on the first big fire I ever witnessed up close. I was a newcomer, wide-eyed at everything about the firefighting world—from the scope of the fires themselves, to the guys on the ground, to the pilots, the tactics, and the equipment. During this first exposure to the aircraft, I was looking at a pair of CL-415s that were assisting us from the OMNRF. It foretold things to come, as I would eventually buy several of these aircraft and source almost all of our company's early scooper talent from the ranks of the OMNRF.

Every type of aircraft had its place, from the Twin Commanders on air attack to the retrofitted tankers dumping retardant. But there was something about the Super Scooper that provoked a feeling akin to awe. It was so strange in appearance—big and chunky and brightly colored, and yet so nimble. It could get so close to the flames, dumping water directly on the fire itself. Over and over. Unlike tankers, the scooper could work all day without having to return to base. All it needed was a nearby body of water. It would swoop in, fill its belly from a lake or river, then dash off to the fire and dump its payload. It seemed like such a smart and effective approach to fighting wildfires, especially on an initial attack when the fires were relatively new and manageable. I didn't know much about the business or tactics at that time, but I remember watching the scoopers and thinking, *Man, someday, we have to get one of those.*

★ TIM SHEEHY ★

It really was that simple. We started with a mission—"What is the best outcome on the ground?"—and worked backwards. I had seen the tankers at work, and I had seen the scoopers at work. The former looked like airplanes. The latter looked like absolute beasts. They looked like Transformers—like they were built specifically to go out there and crush things. Which they were. And the thing they were intended to crush was fire. That's all. That was their *raison d'être*. They were the only purpose-built firefighting plane in the world, and they had demonstrated amazing operational performance and safety.

Of course, I had no idea of the financial and bureaucratic obstacles that would stand in the way of acquiring something that seemed like such a potent weapon in the battle against wildfires.

In the mid-1960s, the Canadair CL-215 amphibian airtanker became the first airplane designed and constructed for the express purpose of aerial firefighting. Previously, every other aircraft that flew on wildfires had been adapted from some other job, including the assortment of aircraft that had been used as "water bombers." The plane's initial design grew out of a December 1963 meeting on forest fire protection held in Ottawa, Canada. Three years passed before a decision was made to begin production of the CL-215, and three more years went by before the plane was delivered to its first customers: the Province of Quebec in Canada and the French Protection Civile, which ordered a combined thirty-five CL-215s. This was not a small order and reflected Canada's commitment to using the new aircraft as the foundation of its forest fire detection and suppression efforts. The planes could also be used in backcountry search and rescue missions and for military and coast guard observation—versatility that made it even more appealing.

Over the course of the next twenty years, a total of 125 Canadair CL-215s rolled off the production line. Although initially intended for fighting wildfires in the vast Canadian wilderness, the CL-215 found an eager market throughout Europe and in parts of South America and Southeast Asia. The appeal was obvious and logical. The CL-215 was outfitted with two water tanks holding a combined 1,400 gallons of water, along with two foam retardant tanks that could be employed as needed. The CL-215 could fly low and slow, dipping to within one hundred feet of a fire before opening the doors on its belly and dropping upwards of ten thousand pounds of liquid directly on the flames. In comparison to air tankers, which sprayed retardant from several hundred feet above the fire, it was a much more precise mode of suppression. Not "better," necessarily, but different—and certainly a more effective tool in the early stages of a fire.

Moreover, the fact that it could "reload" dozens of times in a single day—filling its tanks in just ten seconds while skimming the surface of a body of a lake or river—without having to return to a fire base (as a tanker must do after every drop) made the CL-215 even more attractive. A single CL-215 supposedly completed more than two hundred drops in twenty-four hours while fighting a fire in Yugoslavia. Whether this is true or apocryphal is hard to say, but this much is certain: the scooper had range and ability that no tanker could match. Early skeptics wondered whether the CL-215 would be practical in areas where large bodies of water were not present. But in Canada and most of Europe, such areas were rare and largely unaffected by wildfires. Additionally, the scooper was so agile that it could drink from a lake or river as shallow as six feet.

Ungainly in appearance but dexterous in application, the CL-215 was a wondrous little workhorse made even better with the addition of twin turboprop motors on the CL-215T. Even more impressive and effective is its larger and more modern sibling, the CL-415, introduced in 1994 and equipped with larger water tanks capable of carrying more than 1,600 gallons of water (or water mixed with foam retardant) and four drop doors instead of two to allow additional drops if desired. The CL-415 program was purchased by Bombardier Aerospace, which oversaw production of more than fifty CL-415s (dubbed "Super Scoopers") that were utilized throughout Europe and Canada. Production slowed dramatically in the 2010s, in part because of Bombardier's financial difficulties, and ceased entirely by 2013. Shortly thereafter, Viking Aircraft acquired rights to the CL-415 certificate. Viking would eventually restructure and rebrand under De Havilland Aircraft Canada, which is its status as of this writing.

Although scoopers continued to fight fires around the world, the production of new aircraft stalled partly because two of the biggest potential markets on the planet—the United States and Australia—declined to embrace the technology. The Los Angeles County Fire Department utilized the services of two scoopers starting in 2006, but these were provisional contracts, not purchases. In fact, North Carolina and Minnesota were unique among American states acquiring scoopers outright (though both states employed the CL-215 rather than the newer CL-415).

The reluctance to incorporate the CL-415 into wildfire management left me scratching my head. Here was a seemingly useful tool employed by almost every other first-world country in wildfire management and suppression, but it was virtually ignored—or actively rejected—in the United States. It made no sense. Today, most countries use scoopers as the core of their aerial firefighting fleet. Canada has eighty-six of them flying as of 2023. Europe has more than ninety. In the U.S., however, we had a grand total of four in 2015. We have the most active fire landscape in the world, and yet bureaucracy keeps us wedded to antiquated technology and tactics. It was enormously frustrating and would eventually inspire the writing of this book. As I tried to understand why there was institutional resistance to this amazing tool, by necessity, I ended up learning the historical facts.

In our first few years on the wildfire scene, I would watch the scoopers make periodic visits to the U.S., and I was always impressed by their capabilities. Again, this is not to diminish the contributions of tankers and the people who fly them. The use of tankers and retardant clearly has its place, as does the use of Super Scoopers. It seemed so obvious! So why was there so much pushback whenever I brought it up? The answer: traditions die hard, especially when there is money involved. Retardant is expensive—between ten and thirteen dollars a gallon as of this writing. A tanker will unload anywhere from two thousand to nine thousand gallons per drop, depending on the size of the plane. On a good day, that tanker will do six or seven drops, maybe eight if the fire is near a tanker base. Notice I didn't say "airport." That's because not every airport is equipped with retardant and pumps suitable for refueling. All of this—the retardant, fuel to and from the airport, wages for the crews and ground staff—contributes to the cost of a mission. It's fair to say that every active tanker is burning through hundreds of thousands of dollars per day.

Good intentions and skilled pilots notwithstanding, tankers sometimes miss their mark. They are dealing with difficult weather conditions, constantly shifting fire patterns, and they are dropping from several hundred feet in the air. It's impossible to pitch a perfect game. Unfortunately, retardant that misses the mark is basically useless. There is no recouping that investment; there are no do-overs. Well, there are, but they are costly.

Water on the other hand, is free. If you're a little off the mark in a Super Scooper, you circle around, grab another load, and try again. All in a matter of minutes. As opposed to the tanker, which might have to fly three hours to the nearest base to refuel and gather more retardant. Common sense would seem to indicate that a combination of tankers and scoopers would be the best approach depending on weather conditions, location, and the size of the fire. In countries where scoopers are extensively or primarily used, the focus is on fire suppression as opposed to fire management. Additionally, the resources—that is, aircraft and personnel—are mostly owned and controlled by governmental agencies. This is not the way it works in the United States.

Firefighting aircraft in the U.S. are owned and operated by private companies like Bridger Aerospace. The corporation provides the pilots, maintenance crews, and aircraft through a contract with the U.S. Forest Service or some other governing agency. That agency also pays for the retardant that is pumped into a tanker, which means that the American taxpayers are footing the bill for all this retardant, some of which, incidentally, is provided by companies with a history of shady business practices and government kickbacks. Beyond these ethical lapses, there is a more practical reason that retardant use should be of at least some concern, and that is the toxicity of the chemicals involved. Corporate protestations to the contrary notwithstanding, retardant can be poisonous. And yet, every year, we drop millions of gallons of these chemicals on our most sensitive watersheds and national park lands. I personally am not against the use of retardant; it's a critical aspect of firefighting tactics. However, retardant is merely one weapon in the arsenal, and its use is not without consequences.

In the fall of 2022, an environmental group filed a lawsuit in a Montana federal court alleging that the U.S. Forest Service polluted waterways by inadvertently dropping fire retardant in or near waterways. The retardant was dropped by aircraft under contract with the Forest Service while assisting wildland firefighters on the ground.

The suit, filed by the Forest Service Employees for Environmental Ethics, claimed that government data revealed more than 760,000 gallons of fire retardant was dropped into waterways between 2012 and 2019. The lawsuit alleged the continued use of retardant from aircraft violates the Clean Water Act and requested a judge to declare the pollution illegal. The Forest Service has established retardant avoidance areas along waterways where the liquid is not supposed to be applied, effectively creating buffer zones around waterways and habitat for some threatened species. Approximately 30 percent of USFS lands are restricted from retardant use, with exceptions when human life or well-being is at risk. The policy was the result of a 2010 Environmental Impact Statement that studied the use of retardant and how it affects water resources and certain ecosystems.

This is an enormously complex and politically charged issue. Ever since the early use of borate-based fire retardants, there has been controversy surrounding their environmental impact. In fact, all the different types of chemicals brought to bear against fires in any environment end up being controversial—from asbestos-based fire protection products in buildings and clothing, to foam-based additives in water used for chemical fires, to halon-based fire extinguishing agents, there is no getting around the fact that using anything other than pure water, sand, or fire itself to fight fire will involve a nasty and potentially toxic mix of chemicals. Environmental and worker safety organizations have built a cottage industry around targeting various chemicals (and chemical manufacturers) for all types of fire mitigation. In the case of wildland firefighting, the most common target has been Phos-Chek, a product routinely implemented to slow the spread of wildfires in the U.S. and throughout the world.

The basic argument presented by the FSEEE lawsuit is as follows: fire retardant contains chemicals that are harmful to sensitive riparian areas and when dispensed in these areas, it represents a significant threat to various species and the broader balance of the wildlife in the area; thus, the U.S. Forest Service should cease all retardant dispensing operations. The response of federal agencies, although cumbersome and bureaucratic, essentially admitted that occasionally and unintentionally the government has dropped fire retardant in watersheds, and yes, this is regrettable. But when placed in the context of an emergency response scenario and weighed against the broader danger to the area of a megafire that obliterates the entire ecosystem, this is a risk worth taking.

This latest case brings into focus many competing visions for wildfire management in the United States but most notably illustrates a fundamental divide: *let the fire burn* vs. *fight the fire*. A significant portion of the environmental lobby believes that all active fire suppression efforts are bad, and we must return to a natural fire management construct wherein we allow fires to burn uncontained as that is Mother Nature's way of restoring and rejuvenating the landscape. In a vacuum, this mindset is admirable, and I haven't heard many arguments against it in a theoretical sense. But, like socialism and communism, while the theory *sounds* great, in practice, it often fails rather dramatically.

With fuel loading where it is, some of the current El Nino spells that are in their high-fire cycle, and the wildland urban interface growing at a rapid pace, it's simply not economically responsible or socially acceptable to let these areas burn. Additionally, thanks to massive timber loading, the temperatures of modern wildfires are getting so high that they often don't rejuvenate the landscape; they destroy it. Permanently. Fighting these incidents quickly until prescribed burns and clearing can be employed is the only option for many of these forests. For that to work, we need to be able to fight the fires with retardant, as well as with water and other resources.

I am a proponent of using water whenever and wherever possible. But the outright banning of fire retardant would require a vast overhaul of tactics, techniques, and procedures across all agencies. You would not be able to utilize ground teams as aggressively knowing that they would not have the protection of fire retardant in the case of a rapidly advancing flank. But activist lawsuits are becoming the norm in the United States and there is a lot of money to be made by those pushing this agenda. As of this writing, District Judge Dana Christensen of the U.S. District Court in Missoula has ruled that retardant can continue to be used, but that the USFS must now consider air tankers to be "point sources of pollution," and the agencies must therefore work with the EPA to permit the use of retardant as a hazardous substance. The implications of this could be widespread and potentially impact the willingness of operators, pilots, and aircraft owners to partake in aerial firefighting.

It's important to understand why the use of retardant is so widespread. The U.S. has long employed an indirect attack method of fighting wildfires. When a new fire starts, unless it is extremely small (maybe fifty feet wide), a ground crew will not be dispatched in the same way that an urban firefighting crew would be deployed. The ground crew is not expected to run up to the flames and start spraying water on it and tamping it down with shovels. That's just not practical, nor is it usually possible given how quickly wildfires spread. By the time you get to a fire, you're often dealing with one hundred to two hundred acres of coverage—far more than can be beaten back by ground crews alone. Instead, the ground crews will determine the direction of travel based on terrain and weather conditions. For example, hillside fires naturally burn uphill. If it's windy, the fire will be driven in the direction of the wind. You can't change any of that, but you can get in front of the fire and try to reinforce a natural break—a ridgeline or a creek bed, for example. Ground crews will reinforce the break using chainsaws and shovels, stripping trees of bark, removing vegetation, even utilizing a controlled burn so that when the fire reaches the natural break, it runs out of fuel and dies out.

At the same time, tankers ideally will come in and either enforce that firebreak by dropping retardant on the break or by extending the break if ground crews lack manpower. In effect, they create a giant V-shaped firebreak that serves as a dam when the fire arrives.

That, in a nutshell, is the indirect attack method of fighting fires, and it is how fires have always been handled in the United States. Retardant is not meant to be used like water or a fire extinguisher—directly applied to the flames. It is intended to be dropped around the perimeter of the fire (thus the name of the retardant company Perimeter Solutions), in the hope of containing or managing the fire.

In Canada and much of Europe, they take the opposite approach. There, a direct attack model of fighting wildfires is the preferred model. As soon as a fire starts—any fire!—the goal is to put out the fire as swiftly as possible. And the best way to do that is by dumping tons of water directly on the flames. Obviously, tankers and retardant are used as well, but they are not the primary weapon. It is a dramatically different tactical and philosophical model. And it's worth noting that by far, the biggest and most destructive complex fires on the planet over the last decade have been found in the U.S. and Australia—the only two countries that remain wedded to indirect attack as the primary method of fighting wildfires.

Inertia plays a large role in this. All bureaucracies (and many organizations in general) tend to keep doing things the way they've always been done simply because it's the path of least resistance. Change takes time and effort and, sometimes, money. It also tends to be disruptive of long-standing relationships. It is, in a word, uncomfortable.

But that's not a valid excuse for maintaining the status quo—not when you're facing a foe as formidable as the ever-expanding wildfire season in North America. When I first became enthralled by scooper technology and started inquiring about their availability and asking why they weren't more prevalent in the U.S. firefighting landscape, I got a lot of vague responses about how they weren't practical in many environments (wrong) and how tankers were the backbone of the U.S. firefighting system. They were sturdy and reliable. Most of all, I was told, they were more cost-effective than scoopers. And therein lies the problem: faulty reasoning based on the fact that the U.S. government fights fires through the prism of a daily cost model.

A DC-10 holds approximately nine thousand gallons of retardant. Even smaller tankers can carry three thousand gallons, which is twice what the Super Scooper can carry. As noted, however, the scooper ends up dropping far more payload on a fire over the course of a day because of its ability to effortlessly reload multiple times, so you can certainly argue that it is a more cost-effective way of doing business, especially when you factor in the cost of retardant and fuel. But—and this has been a significant hurdle for the scooper and its advocates to overcome—the Super Scooper is a more expensive aircraft to purchase than any tanker. By a wide margin. A new CL-415 costs roughly $30 million—twice what the average tanker will cost. A company that purchases a Super Scooper must find a way to recoup that investment through federal or state contracts, and there has been a reluctance among those entities (the U.S. Forest Service, CAL FIRE, and others) to support the venture. As a result, they just don't show up very often on the American wildfire scene. And when they do, they are typically on loan from the Canadian government, and only for a brief period.

By our third season of fighting fires, I was completely enamored of the functionality of the Super Scooper and became obsessed with trying to acquire one for Bridger Aerospace. Part of this was business-related—I honestly thought it made financial sense to expand our fleet with scoopers since they were so scarce in the U.S. But I was mainly drawn to the capabilities of the scooper and what a fantastic yet underutilized weapon it was. Not only did it make sense purely from a tactical standpoint to complement indirect attack with direct attack, but it was an environmentally friendly approach to firefighting, as well. In light of all the concerns with retardant, why soak the ground with chemicals if it's not necessary? Why not use water wherever it's readily available? There is a cost to using retardant, and it's not just monetary. Every ounce of retardant must go somewhere; it doesn't just miraculously disappear. Even if it does its job and helps stop a fire, it winds up draining into waterways or groundwater. It can poison fish and other wildlife, and it taints potable water in surrounding communities. It can also be harmful if the vapor is ingested as the retardant burns.

If this sounds like a contradiction to what I said earlier, well, let me clearly and emphatically repeat: retardant is an invaluable tool in the fight against wildfires and is a critical aspect of fire strategy and containment. However, there must be a balance to achieve the most successful result. Water-based direct attack is a critical piece of any aerial firefighting strategy. Tankers and retardant have their place in the fight against wildfires, but if there are other less harmful options—if there are cleaner, more efficient options—then why not use them?

For better or worse, the tanker model has had a stranglehold on the American wildland firefighting industry for a half century, and the industry is understandably reluctant to loosen its grip. The current methodology has served the companies that own and produce the tankers well, and it has encouraged complacency within regulatory agencies. Ours is a tanker-centric industry. No one knows any different, nor are the old dogs particularly eager to learn new tricks. Meanwhile, the rest of the world adapts and changes and trics to keep pace with the broadening threat of wildfires by any means necessary. And that includes, primarily but not exclusively, utilizing direct attack protocols with scoopers on the front line.

★ TIM SHEEHY ★

And you know why? Because it works! You don't see as many news reports about towns in Canada burning up. Or Spain. Or France. Yes, they have fires—of increasing scope and intensity, just like almost every other country in the world. But they have a national posture where they treat firefighting as an emergency. They fund it and manage it like the military, so they buy the right planes to do the job, they enlist professionals to manage and operate the technology, and they crush the fires with expediency. And water. Lots of water. Retardant alone simply isn't enough. I would argue that it shouldn't even be the cornerstone of firefighting (except in cases where a controlled burn is logical and safe—options that are becoming more problematic thanks to fuel load, climate change, and civilian encroachment on wildlands). Too often, when only retardant is used, a fire will jump the firebreak created by retardant, thus starting a new fire on the other side of the line and wrecking a crew's entire containment strategy.

Whenever possible, start with water. Then add more water. And keep adding water (along with retardant) until the fire is out or manageable. Each large air tanker typically drops one load of retardant every two to four hours (with travel and reloading time), so even if you have multiple tankers on a fire, there are still sizable chunks of time when no retardant is available; thus, a combination strategy works best. Bring in the scoopers—even in a complex fire that cannot simply be snuffed out, scoopers can greatly aid the effort by repeatedly dumping water and cooling the fire (sometimes in as little as three-minute intervals between drops), slowing it down a bit, thus giving ground crews time to reinforce the line. If all goes well, particularly when scoopers are used in direct attack, before the fire has exploded, you might just extinguish it entirely. I've seen plenty of occasions where this has happened: scoopers basically dousing a fire like a rainstorm. It's hard to watch the scoopers in action and not be impressed. At the very least, it's hard to argue that we don't need more or that they should merely be supplemental tools. I understand territoriality and the natural inclination to continue on a given path simply because you've been on it for fifty years. But the threat of wildfires is not going away, and we need to use every tool at our disposal. You'd never see a carpenter show up at a construction site with just a hammer or a drill. Certainly, you'd never see him turn down a tool that will make his job easier and more efficient. And safer. It takes multiple tools to do any job properly. And even as a neophyte, I could see that the American wildfire toolkit was missing a few items that should have been readily available.

It made no sense.

"What I see in America is basically what I saw in Canada thirty, thirty-five years ago," said Al Hymers. "Boots on the ground first, aerial support second. The CL-215 and then the 415 changed all of that in Canada, and I think it's changing things in the U.S. Of course, in Canada, there's water everywhere, and that isn't necessarily the case in America, so you have to take that into consideration. But there are plenty of places where water is readily available, and the scooper does a hell of a job in those scenarios."

What I see in aerial firefighting today is not unlike the U.S. Navy in the 1930s, when the old guard clung to the "capital ship" mentality, wherein the entire fleet revolved around the venerable and powerful battleship. For hundreds of years, "ships of the line," as they were called, served as the foundation of western military fleets around the world. Empires had risen and fallen based on the outcome of brutal naval battles that amounted to opposing warships facing off—*in a line*—and firing at one another. When new technology became available to enhance the range and accuracy of naval gunfire, the admirals of the U.S. Navy dug in to make their battleships the newest and best in the world.

As World War II approached, a new generation of naval leaders challenged the battleship community. They understood the disruptive effects of submarines and aviation and how they would upend centuries of naval doctrine. Most notably, aircraft carriers would create a platform to project power over an entire region of ocean, far greater than the distance of fourteen-inch guns. Exercise after exercise reinforced the notion that aviation was going to reshape all future naval engagements and largely relegate the battleship to a supporting role. But casting aside five hundred years of naval doctrine was not easy for U.S. Navy brass. Many of them dug in and refused to change. Until a bold, carrier-based aviation raid on Pearl, coupled with the amazingly swift conquest of Southeast Asia by the Japanese war machine, crippled America's Pacific forces. Not long after that, the die was cast and the U.S. Navy became structured entirely around the Carrier Battle Group, a doctrine that persists to this day—at least until the next major disruption makes change inevitable.

Bridger was only a couple years old when we began looking hard into the possibility of expanding our business into suppression aircraft, which meant one of two things: tankers or scoopers. I had been talking extensively with my brother, the financial guy in our organization, and told him that if we could find a way to incorporate scoopers into our lineup, we might really have something special—a service that was not only effective, but that no other U.S. company offered.

"I don't know shit about financing," I said to Matt. "But I've seen these things in the field, and they are amazing."

"Okay, you find a way to get them, and I will find a way to pay for them."

Theoretically, it would have been easier to get into the tanker business since a well-trod route had already been established. But it was not a route without obstacles, particularly if you wanted to start from scratch. Buying a tanker was one thing—and came with its own laundry list of headaches—but buying a commercial 727 and converting it to firefighting capability was a logistical, engineering, economic, and bureaucratic nightmare. The airworthiness and oversight process was endless. You started out knee-deep in shit and it got deeper every day. And at the end of the journey, you ended up with something already offered by several other companies. It wasn't that air tankers weren't effective; it was just that the nation already had enough of them. I had talked to enough firefighters on the front lines to know what they wanted: more resources devoted to initial attack and direct attack—the tools to *suppress* fires, not simply manage them—because too often, they had become unmanageable. There was, it seemed, a market gap, and we decided to try to fill that gap with Super Scoopers.

As I examined the history of air tanker conversion, I started poking around in 2017, making calls to determine the viability of our plan. What I soon discovered was that while there were plenty of Super Scoopers flying throughout the world, the CL-415 was no longer in production, and not a single new aircraft had been manufactured in nearly a decade. Viking had recently acquired the CL-415 technology from Bombardier, but when I reached out to Viking to inquire about buying a scooper, I was told, "Well, we're not really in the business of building airplanes." This was true. Viking's mission was (and remains) keeping old Canadian aircraft flying through the acquisition and support of Canadian-built aircraft produced years—sometimes decades—earlier by companies that were no longer in business or at least no longer manufacturing these aircraft. Specifically, Viking today is the Original Equipment Manufacturer (OEM) and Type Certificate holder for all out-of-production De Havilland Canada aircraft and the Canadair Amphibious Aerial Firefighting aircraft fleet.

The Super Scooper fell under this umbrella.

While trying to track down the CEO of Viking at a sales conference, I ended up talking with one of their sales reps for a while about the business, and I told him I understood that Viking was mainly a supplier of parts. Nevertheless, we were exploring the possibility of perhaps buying a scooper someday, and I wanted to get his thoughts on the process. Since Viking owned the certificate for the CL-415, was there any possibility that Viking might want to restart the program on some sort of limited scale?

The guy responded by literally laughing in my face.

"Come on, man. You guys could never afford that. And anyway, we only sell to government agencies."

He walked away and left me standing there, briefcase in one hand, bottle of water in the other. I was too young and dumb to even be offended, so I just sort of shrugged, said, "Huh, okay," to no one in particular, and went about my business. Eventually I met with Rob Mauracher, who was then COO of Viking and repeated my interest in incorporating Super Scoopers into the Bridger lineup. Rob was much nicer, but his response was essentially the same as that of his sales guy: "These things are priced for a government market."

I nodded. "Yes, I understand. But I still want to buy one. Maybe more than one."

Now, I was absolutely nobody at this time, and Bridger was still the new kid on the block. Sometimes that makes it easier to put yourself out there and take risks without worrying about what anyone else thinks. Rob was neither encouraging nor discouraging, but rather just sort of neutral on the subject—he likely figured I'd just go away when faced with the economic reality of the situation. Instead, this turned out to be just the first of many discussions we had over the next few months, an evolving dialogue about how we could restart the scooper program. Building brand new aircraft from scratch with new design configurations would have been prohibitive from both a cost and time standpoint; instead, Viking proposed a modernization of the CL-415 (itself a modernization of the CL-215). There was just one catch: Viking wanted another customer as part of the deal and a minimum order of three aircraft, one of which would be reserved for Bridger.

"Fair enough," I said.

Any big business deal will die a thousand deaths, and that was certainly true with the resurrection of the Viking Super Scooper program and its partnership with Bridger Aerospace. Raising capital is hard and mostly awful and soul-crushing, but it's a necessary part of growing a business. And finding someone to fund an almost $200 million order of out-of-production aircraft, to be operated by a three-year-old company, led by a nobody, for a contract that didn't exist, with an agency that couldn't always agree on the required specifications for their aircraft, all while fitting into the asinine Small Business Administration guidelines, became a gargantuan undertaking on its own. As I discovered throughout this process, it's almost impossible to enter the U.S. aerial firefighting market unless you are already in it or are an exceptionally wealthy family or individual.

My brother and our finance lead, McAndrew Rudisill, eventually found our partner in the Blackstone Group. The CEO of Blackstone, Stephen Schwartzman, is one of the greatest business leaders in the history of the United States. A hard-working visionary and a former Army officer, he also understands the horsepower of our veteran community. He had once seen the Super Scoopers operating in France and was immediately impressed with their capabilities, like I was. When he heard about our deal, he was very excited to proceed and provided us with a crack team at Blackstone to execute the deal. Wayne Berman, Todd Hirsch, and David Kaden, all incredibly accomplished men in their own right, joined our board and helped us get a deal together. These are partners and friends we still have to this day.

But even after we acquired the necessary funding from the Blackstone Group, the waiting continued. Communications from Viking had gone eerily silent, and I was worried our deal was falling apart. This was the future of the business and after all the effort to get it to this point, I wasn't going to just let it slide because my emails and phone calls weren't being returned. I found out where the Viking executives would be—time and place. No better way to communicate than in person. So I hopped on a plane and flew to Chile to track down the Viking team at an airshow—he'd been dodging my calls and emails for weeks. So, without any guarantee that I'd find him or get to speak with him, I bought a last-minute ticket in coach and flew for twelve hours in a middle seat. I rented a car and drove to the show. Although I had been fluent in Spanish many years earlier and had used it a bit in South America during some military operations, I had to unexpectedly break it out again in the streets of Santiago. Fortunately, it all came back to me.

I found Rob at his booth, and he actually seemed happy to see me.

"So, where's our deal?" I asked.

"You tell me; we are ready to approve. But we're concerned about your funding status. You still haven't given us the comfort we need to know this is a legitimate order."

"What do you need?"

"Let us meet your financiers. If it's good, we will sign and get to work."

"Understood. I'll make it happen."

It was a long trip to Chile to have that discussion. But that's business—sometimes you just have to do it in person.

Three weeks later, we were in Calgary with our partner and champion from Blackstone, David Kaden. David had really become part of our team after we secured a reasonable path forward with Viking early in 2018, and was critical in much of our success.

We ultimately agreed to order three scoopers at a cost of $30 million per aircraft. The down payment was $1 million per plane, with the remaining balance due when we took possession of the aircraft. But for the longest time, it was all sort of theoretical. Our down payment was refundable if Viking did not go through with production. Viking, in turn, was not going to authorize production until it was certain we had secured funding for the entire $90 million and it had found a second customer to supplement our order of three aircraft. Preferably one with a bigger name and résumé than Bridger. Viking was reluctant to move forward with the project until it had orders for a total of six planes. We represented half of that allotment but comprised the entire order at that time.

Every day, it seemed, we were on a treadmill, going nowhere fast. And there was nothing we could do about it.

I learned early in my career in the military that minimizing variables is a critical capability of a leader. Figure out what you can and can't control. Minimize or limit as many variables as you can. Just get rid of them. It's a lot easier to come up with a plan if you have a reasonably good idea of what to expect. It's like playing poker: the more cards you know your opponent has, the stronger your position gets. So, I went to Viking and said, "Listen, it's been months now, and you guys still haven't sold the other three planes. I need to get this program kicked off because now I have the funding to launch it, and my partners are eager to go to get going. They want to put their money to work. But I can't go get business—I can't secure contracts for these planes—until you start making airplanes."

And they said, "Well, we're not going to make airplanes just for you. We told you that—we need other more established customers to de-risk the program."

Okay, so how could I eliminate or minimize this variable? Simple.

"Let me buy four planes. Then you guys only have to sell the other two."

"Deal!"

More time went by without any other customers stepping up to the plate or scoopers going into production. Either I was some sort of visionary or the stupidest guy in aviation. Optimism turned to pessimism. Frustration grew.

"What if buy a fifth plane?" I suggested. "Then you only have to sell one more."

"Okay, fine."

More time passed. Still no other customer, still no planes. Our investment partners understandably grew restless. *Maybe*, I thought, *we should have invested in tankers after all. Stop swimming upstream.*

Late in the summer of 2018, I went back to Viking and put all my remaining cards on the table.

"So what's going on? Are you guys going to launch this program or not? I am trying to build a business around your planes. Are you just stringing me along?"

This was vigorously denied.

"It doesn't matter," I said. "We can't wait any longer. Let's make this simple. I'll buy all six planes."

"I'm not sure we are comfortable taking that risk as a business."

★ TIM SHEEHY ★

"Well, now it looks like I'm the one taking the risk on you. I have been steady in my resolve; you guys still haven't committed. You said you need six to get your program started, so…let's get started. We have the money, we're serious about this."

"Okay. We need to take it to the board, but we believe in you, and we want to get the program going. We will de-risk ourselves on the remaining six airframes."

This time, we really did have a deal. Bridger Aerospace became the sole customer for the relaunch of the Viking Super Scooper program with an order for six planes. I found this to be both thrilling and astonishing. Just three years earlier, we were a tiny company with three small planes and less than a half dozen employees. And now we were responsible for bringing back one of the most iconic aircraft in the firefighting industry. We had literally brought it back from the dead—this plane that was a symbol of Canadian heritage and had become one of the most prolific and widely disseminated firefighting aircraft in history. For half a century, it had been the only aircraft in the world designed for the express purpose of fighting wildfires; that it had apparently become expendable was, to me, a tragedy. I had talked with enough firefighters and pilots. And I had seen enough destruction to know that we needed more scoopers in aerial firefighting, not less. And yet, for some reason, there was resistance.

"Here's a good example," remembered Hymers, talking about an incident in 2019, when he and several other Bridger pilots attended a National Aerial Firefighting Academy conference. "It's one of those mandatory events where everybody is telling you what needs to be done and how it should be done. At one point, we have these breakout groups, and the instructor went around the room asking everybody about their background. "Who here flies the large air tankers?' A couple guys put their hands up. 'Who flies the large helicopters?' One or two guys raise their hands. 'Who flies air attack?' Two or three more raise their hands. On it goes, all the way down to the bottom. Just before he asks who sweeps the hangar floor, I get the guy's attention and say,

'What about scoopers?' And the guy just laughs, shakes his head, and says, 'Yeah, scoopers. Right.' And I thought, *Why, you son of a bitch. You have no idea.* But that's the way we were treated.

"On the first day of the conference, we even got a call from Tim Cherwin asking us to tone it down a bit because he'd gotten a complaint from someone at the Forest Service. 'Those pilots you got,' the guy had said. 'They do realize they're working in America now, right? They're not in Canada anymore.' And, you know, we weren't trying to be rude. But people were telling us how to do our job and not listening to our input, even though we might actually have a better idea than they do."

Added Jason Robinson of AeroFlite: "For a while, scoopers were the red-headed stepchildren of aerial firefighting [in the U.S.]. But I do think that perception is beginning to change. Part of the problem is the way it's been marketed to the firefighting community. If you walk into a room and tell everyone that your product is the greatest thing since sliced bread, they probably aren't going to react well. You need some humility.

"We've gained acceptance not because we've got a new airplane, but because we've altered the way we try to fit in and operate. It's an attitude of being supportive of the ground crews and coordinating and building trust with other air resources. The biggest compliment I can get is when we go to get fuel and that Chinook pilot or a tanker pilot says, 'Hey, Scooper 261, nice working with you.' And I'll say, 'Yep, see you tomorrow.' That's a great feeling. Whereas there were many times when I first started where there was an adversarial thing on the radio—where helicopters felt like maybe we were taking their work and they didn't want to talk with us."

★ ★ ★

By March 2020, we were preparing for the test flight of our first scooper, tail number SN 1081. We were on an aggressive schedule, and it was going to take everything going perfectly over the next several weeks for us to be ready for fire season. Viking would have to finish the planes, Transport Canada was going to have to be spot-on with their work to get the airworthiness certificate and type certificate data sheet shared with FAA, the FAA was going to have to move fast to accept it, and then the USFS was going to have to be ready to pounce on the plane, inspect it, card it, add it to the contract, and then we were going to get to work. It was going to be a herculean feat of bureaucratic Judo to get this done in time for fire season, but I was confident we would be able to manage the challenge. And then…the world stopped turning.

I was with Darren Wilkins, our COO; Andrew Hill, our head of Scooper Maintenance; our first two Scooper mechanics; and Hank Williams, one of our top pilots, in Abbotsford, British Columbia, on March 14, 2020. The reports of COVID around the world had been swirling for a couple of weeks at this point, but it was still a bit uncertain. Was this another SARS scare? A regional virus that would be restricted to Asia? Swine Flu? It was unclear what the severity was, but one thing was certain: none of us had any idea what was about to happen to the world, our lives, or our business as we flew up to Canada for this trip. We were on the hangar floor inspecting the airframe on March 13 as it started to become clear that this was going to be a very serious global issue but would probably pass in a few weeks. What was more concerning was that a nursing home in Seattle was becoming the U.S. epicenter of the virus and was causing an unprecedented "lockdown" response in the Puget Sound area. Abbotsford is just across the border from Seattle, so we started worrying that we might end up being included in this radius.

When we came back to the hangar on March 14 and continued our inspection, we began to see news stories of an impending border closure. We also were hearing that U.S. ports of entry were going to start closing. We had planned several more days of inspections, but it became clear that with uncertainty mounting and a lot to do back home, we had better boogie out of Canada. We immediately had our Twin Commander pulled out of the hangar and started to file our flight plan home. Our worst fears were beginning to materialize. We were being denied our flight path back to Montana by Customs and Border Protection (CBP). We called the CBP center for our region and asked why we couldn't get back to Montana.

"Neither Great Falls nor Bozeman are approved entry points for international flights anymore," we were told. "You have to go to an approved COVID screening facility to enter the U.S."

"Okay, then, where do we go for that?"

"SEATAC."

"SEATAC? That's where the outbreak is right now! In order to screen us for the virus, you want us to fly to the only airport in the U.S. that is actually at the center of this thing, and potentially spread it to Montana from there?" Wilkins inquired into the phone. "That's literally the dumbest thing to do. You want us to fly into the outbreak zone to get screened for the very virus we are trying to prevent from spreading."

"That's the policy. Or you can stay in Canada."

Minutes later, we were loaded up and ripping off the runway from Abbotsford, making the gentle left hand turn on departure to avoid the snowcapped mountains to the east. As we settled into our quick half hour cruise over the Prince William Sound headed south to SEATAC, the usual banter that characterized our company flights was gone. We were all processing the reality of what was going to end up being a very big deal for the entire world. As I stared down at one of the Black Ball ferries crossing from Vancouver, cruising across the sound, I realized we were bearing witness to yet another pivotal moment in history. This was going to be a big deal; it was the same feeling I had seeing the second plane hit the World Trade Center in 2001.

We ended up being the only airborne aircraft in SEATAC airspace. It was incredibly eerie being this small Twin Commander coming in for a landing, seeing all the big airliners parallel parked at SEATAC because they were stopping all movement to arrest the spread of the virus. We landed and were told to pull up to spot S19. I replied, "I will need progressive taxi, I don't know where S19 is."

"Turn right and taxi straight ahead. S19 is off to your left. As we taxied, we quickly realized S19 wasn't a spot—they were sending us to an airline gate. We pulled up to S19 between a 737 at S21 and a 767 at S17. I will never forget the people in the terminal looking down at us through the windows wondering what the hell this little plane with FIRE written all over it was doing at their gate!

The customs agent walked out of the baggage door.

"What are you guys doing here?"

"They told us this is the only port of entry we could clear in."

"That doesn't make any sense. Where are you going?"

"Montana."

"Well, you better get the hell out of here before they shut everything down. You might get stuck here," the customs agent said. Stamp, stamp, stamp. "Good luck."

We didn't need to hear anything more. We got in, fired up, and blew out of SEATAC as fast as we could. As I climbed out over the Cascade Range, I started to realize that this was going to be a huge disruption to our young and fragile business—right when we needed everything in our favor. I had no idea how the next twenty-four months would test the mettle of our business, our customers, our people, our partners, and of course, myself.

★ ★ ★

In April 2020, the first Viking CL-415EAF "Enhanced Aerial Firefighter" was delivered to Bridger Aerospace's facility in Bozeman as part of a contract that, with all options exercised, was valued at $204 million, covering the purchase of six of the amphibious scooping air tankers. Manufacturer's serial number (MSN) 1081, the first Canadair CL-415 to undergo the major modification to the EAF configuration, had taken its inaugural flight one month earlier outside of program-collaborator Cascade Aerospace's facility in Abbotsford. It was a gorgeous aircraft, big and strong, boldly painted in bright red and yellow. The day it arrived in Bozeman, against all odds, at the dawn of a global pandemic, was among the proudest of my life, and certainly the proudest of my business or firefighting career. Our team had rescued an iconic aircraft, one whose widespread implementation could help change the course of aerial firefighting in the United States. We couldn't wait to put it in use, and we looked forward to the arrival of its siblings. This was important. This was progress.

CHAPTER 11

ACCIDENTS WILL HAPPEN

Winter Haven, Florida
February 23, 2019

It was a good day to fly: cloudless, clear, calm. Perfect.

I'd flown in two nights earlier from Milan, Italy, where I'd been attending an operators' conference about Super Scoopers. I arrived around 7:30 p.m. and went straight to bed, knowing I'd be up at dawn for a couple long days of instruction and training, along with a half dozen other pilots from Bridger. Now that we were well down the road to acquiring scoopers and putting them into use as part of the Bridger fleet, there was the small matter of meeting FAA requirements for flying the beasts. The scoopers were multiengine seaplanes, and in order to fly them legally (and safely, of course), a pilot had to obtain a multiengine seaplane rating. This probably sounds simple enough; unfortunately, there were very few pilots in the United States who held a multiengine seaplane rating, in part because there was little use for it, and thus a tiny number of instructors or schools where a pilot could obtain a multiengine seaplane rating.

As we began to investigate the acquisition of the scoopers, our original plan was to assemble a group of pilots who already held a multiengine seaplane rating. Most (if not all) of them would come from Canada simply because there is no seaplane community to speak of in the U.S. Well, there is a small one in Alaska, but it is almost entirely comprised of bush pilots who fly single-engine aircraft. In Canada, where scoopers have been utilized to fight wildfires for many years, and where the Twin Otter, a robust twin-engine nineteen-passenger utility plane, is frequently employed in the bush world, there is an abundance of pilots who hold the multiengine seaplane rating. In fact, a lot of Canadian pilots fly multiengine seaplanes year-round—fighting fires in scoopers for six months and ferrying tourists to island resorts in the Maldives or the Seychelles in Twin Otters the rest of the year. These are solid pilots with years of experience flying multiengine seaplanes in a variety of conditions.

The plan was to hire several Canadian pilots to fly in the left seat of our scoopers (the captain's position) while filling the right seat (first officer) with Bridger pilots who had been flying air attack in the Twin Commanders. But even the first officers would need a multiengine seaplane rating while they were apprenticing, so while hiring experienced Canadian pilots was a logical strategy, it was more of a Band-Aid than a solution. As with so many challenges in building an aviation company, you buy time until you can figure things out. It was only after I began investigating the certification process that I came to realize just how rare it was for an American to hold a multiengine seaplane rating and how few avenues there were to acquire that certification. This should not have been a surprise. If only one half of 1 percent of all pilots have a multiengine seaplane rating, then it stands to reason that there aren't a lot of places to obtain that rating—you can't just go to any flight school and request training that leads to a multiengine seaplane rating because 99 percent of schools do not have a multiengine seaplane in their fleet.

But our team found one in Florida—a small flight instruction business operated by an Air Force veteran and former commercial and private airline pilot named Jim Wagner. Nice guy, sixty-four years old, a legend in the Florida seaplane community, and a highly capable instructor with a ton of teaching experience. The owner of the Twin Bee plane we were training with was unable to fly with us for the day due to surgery, so Jim was filling in. The Twin Bee was a descendant of a single-engine seaplane called the Seabee developed in the 1940s by Republic Air. The Republic Seabee was a rather heavy, bulbous amphibious aircraft that went into production near the end of World War II and quickly became a popular bush plane and air ambulance in countries with long coastlines or a multitude of lakes. Plump and ungainly in appearance, the Republic Seabee was said to resemble a bumblebee in flight—thus the name "Seabee." (It helped that many of them were painted yellow, although that may have been a result of the name more than a cause.) In the 1960s, United Consultants Corporation developed a twin-engine version of the Seabee, which unsurprisingly became known as the Twin Bee.

The original Seabee was a nice plane, flew well, and was much admired for what it did. The twin-engine version was not the upgrade that it was purported to be, but more of a fast-tracked modification that managed to circumvent some of the FAA's more rigorous testing, resulting in a piston–driven plane that was slow, heavy, and neither as safe nor as reliable as its predecessor. The scoopers we had targeted, like most modern aircraft, were driven by turbine engines, which are smaller, lighter, and more powerful. In short, the scooper was like a jet, while the Twin Bee was like a VW Microbus with wings.

There are two reasons for aircraft to employ multiengine technology. The first reason is power. Theoretically, the more power an aircraft has, the more it can carry in both payload and structure. The bigger the plane, the more power it requires to remain in the air and reach its destination. Unfortunately, piston engines are so big and cumbersome that it's often been said that adding a second engine barely accounts for the weight of the engine itself; it's not a loss in power, necessarily, but it's not much of a gain, either. The other reason for adding a second engine is to ensure that the aircraft will remain aloft even if there is a problem with the first engine. It makes sense, right? You would expect that if a twin-engine plane loses power in one of its engines, the second engine would take over and allow the pilot to find a place to land safely. In reality, single-engine planes that lose power tend to fare better in catastrophic scenarios than twin-engine planes.

★ MUDSLINGERS ★

Engine failure typically occurs on takeoff or climb, and when that happens, the single-engine plane will simply not be able to climb through the drag. The pilot will instantly try to bring the plane back down. It's not an ideal scenario, but it is survivable. Modern multiengine turbine-powered planes, like a 757 commercial jet or a super scooper, have more thrust and can easily withstand the loss of an engine. If you lose an engine in a scooper, you just feather the bad engine that went down, and you can fly all day long. Airliners? Same thing. If you're on a climb-out with a 757 losing an engine, it's no big deal. As a passenger, you probably wouldn't even notice. But in older piston–driven propeller planes—put into service before that level of design had been achieved—the loss of an engine was often devastating, resulting in what is known as a Vmc roll. Vmc (V equals velocity; mc equals minimum controllable) is the minimum speed at which a multiengine aircraft can remain in control in the event of a single-engine failure. When an engine malfunctions, thrust is no longer symmetrical, and the plane will begin to yaw and roll. Depending on the type of aircraft, a pilot can maintain control using the rudder and aileron to maintain level flight and compensate for the engine loss. But multiengine aircraft require a minimum speed to keep the plane aloft. If you fall below that speed following an engine failure, the power from the good engine can overpower the air going over the airfoils of the aircraft and flip the plane over, and the plane literally dives nose first into the ground.

For rather obvious reasons, this is also known as a "death roll."

Every pilot knows about the death roll. Since multiengine piston aircraft have largely (but not entirely) been phased out and replaced by more modern turbine-driven aircraft, hardly anyone experiences it anymore, which is good, since, as the name indicates, almost no one survives the death roll. That morning in Winter Haven, I became one of the lucky ones.

I knew we'd be flying a Twin Bee that day, and while it would not have been my first choice for the certification process, I wasn't really worried about it, even after I saw the plane and realized what an old dog it was. We really had no other option. If you want to get a multiengine seaplane rating in the U.S., you take the first opportunity and make the best of it. While this wasn't a big formal school, the instructor had a good reputation and seemed like a nice enough gentleman. So, we all studied the manuals, sat through some pre-flight instruction, did an extensive walk-through, and then began actual flight training. On the first day, only Tim and I would be flying. The other guys were scheduled to fly the following day, so they were all visiting the Kennedy Space Center on Merritt Island some ninety miles away.

It's important to note that we weren't trying to achieve any level of expertise on this particular plane. We were simply going through the certification process for flying a multiengine seaplane. Checking the box, you might say. The Twin Bee was merely the avenue to achieving that goal. In a few months, we all planned to begin intensive training on scoopers, which are substantially better aircraft than a typical Twin Bee. It's fair to say that none of us expected to fly a Twin Bee again.

Although each of us was an experienced pilot, on this trip we were all students. Much like a driver's education class, the seats were reversed: student on the left (typically the captain's seat), instructor on the right. The Twin Bee had room for a third person in the back, watching and taking notes or whatever. That is the seat I occupied on the first instructional flight of the day. Tim sat in the student's seat. It was an uneventful trip, just as you'd like it to be. A little less than an hour and a half of flying in perfect weather. The plane itself was a stubby little aircraft, loud and ungainly, exactly what you'd expect from an old Twin Bee, but otherwise not notable in any way. It was not the kind of plane I would choose to fly professionally or recreationally, but there was no indication that there was anything wrong with it. It was what it was: a half-century-old, piston-drive, twin-engine plane. Like all Twin Bees, it had the appearance of something that was not aerodynamically suited for flight, but it had clearly gotten the job done for a number of years, and Jim seemed totally comfortable in his role as instructor. Based on Tim's flight, it appeared that we were in for a smooth day.

★ MUDSLINGERS ★

On the next flight, Tim and I were supposed to swap places, but he was suffering from a painfully inflamed hemorrhoidal condition (I know—it sounds like a joke, but to a pilot who spends hours sitting on his butt all day, it's an occupational hazard and not even slightly funny), so he opted out of that session to recuperate for a few minutes. I went outside and told Jim there would be only two of us on this flight. He nodded, said no problem, and we strapped in and began going over the flight profile. The optimal time for a training flight is generally considered to be between 1.2 and 1.4 hours. Research has repeatedly shown that after 1.4 hours, the student starts to become fatigued and the absorption of information is impaired. This is true for most types of intense learning scenarios—after a certain period, the sponge is full and the student stops learning. Worse, if the training session exceeds the optimal time limit, the student can develop bad habits that will then have to be "unlearned." I am also a certified flight instructor, and I have seen students hit the proverbial wall, so I always try to stay within the optimal range when giving a lesson—preferably on the lower end to make sure the student is fresh and eager even when the training flight ends. If it's a particularly rigorous flight—with lots of spins and stalls and other demanding (or potentially dangerous) maneuvers, I'll keep it even shorter—maybe slightly under one hour.

Our flight was of moderate difficulty, nothing out of the ordinary. I knew in advance that the first obstacle thrown my way would be to calmly work through a simulated engine failure on takeoff. It's a reasonably benign maneuver, routinely required of a student who is seeking a multiengine rating. To be clear: this was a simulation. At no point in the exercise was the engine supposed to fail or be turned off. It's comparable to taking your foot off the gas in a car—you lose propulsion (the car slows down), and there is the *sense* of losing power. But the engine does not stall or cease to function (as it would if you turned off the ignition). To resume normal speed, you simply apply pressure to the gas pedal.

The aviation equivalent, for the purpose of this exercise, involves pulling back on the throttle to give the sense of losing power. The instructor then tells the student something like, "Okay, we just lost power to one engine. Respond accordingly." The instructor does not—*ever*—intentionally cut power to one of the engines. He throttles back, creating the sensation of lost propulsion, and guides the student through the process of keeping the plane level and maintaining speed under adverse conditions. If, at any time, the student appears to struggle, the instructor uses the throttle to resume normal speed.

Now, there are several modern FAA-imposed guidelines in place to ensure that this process is as safe and smooth as possible. Chief among them is a recommendation that the simulated engine failure take place at a minimum altitude of five hundred feet and preferably more than one thousand feet. The rationale for this is probably obvious: to give the instructor and student time to recover if something goes wrong. But note the terminology used here: "guidelines" and "recommendations." There is no unbreakable rule when it comes to a simulated engine failure; rather, common sense is supposed to prevail based on the accumulated history of such instruction and the accidents that have occurred. An altitude of one thousand feet gives a pilot time to regain control of the plane and to avoid heavily populated areas on the ground if a forced landing becomes necessary. In the case of Winter Haven, Florida, it also gives you time to reach one of the area's many bodies of water (the region's abundance of lakes is one reason it's an ideal area for seaplane training).

When following FAA guidelines, the instructor typically waits until the plane has reached an altitude of one thousand feet before throttling back, and then directs the student to use one thousand as the base altitude. In effect, that becomes a simulated version of ground level, and the objective is to correct the engine "failure" before the plane descends to base altitude. Failure to do so would represent a simulated crash. There are, however, some pilots and instructors who grew up in the sixties or seventies and went through a more rigorous (and, frankly, dangerous) sort of training. If you were simulating an engine failure in those days, you generally didn't wait until five hundred or one thousand feet, you did it right away because that's most likely what would happen in the real world, and there is nothing quite like real-world training. Pretending base altitude is the ground is not quite like seeing actual terra firma rushing up to greet you. I don't disagree with any of that. I was a SEAL and a Ranger. Tough training is usually the best training. But it is not without risks.

I don't know if Jim was a disciple of old-school instruction. I don't know whether he was simply trying to pack as much practical training as possible into a tight window. I also cannot attest to the integrity of the plane. I will say this, and it is a problem throughout the aviation industry: it was an older plane. The fact is, most planes in service today were manufactured in the 1970s and 1980s. That's primarily because Part 23 of the FAA's Code of Federal Regulations governing airworthiness has made the development of new and genuinely innovative modern aircraft cost prohibitive in most cases. There are so many hoops to jump through and so many legal and regulatory costs (in addition to already stratospheric engineering, design, and production costs). That's why the 737 hasn't really changed much since it was introduced in 1967—it's much more economically viable for a company to update an older model of aircraft than it is to develop a new aircraft entirely from scratch. Thus, older planes remain in circulation for decades, and when it is time to retire them, they usually are replaced by a newer version of the same model, designs for which are already in place, with full FAA support and approval. Unless you are part of the aviation industry, you're probably blissfully unaware of the age of most fleets (and despite this fact, commercial air travel remains one of the safest modes of travel because it so tightly regulated, with every conceivable safety precaution in place). But imagine if you were still using the same Motorola brick phone you were using a quarter century ago. Imagine if you were still driving a 1975 Cadillac. Not a used '75, mind you, but exactly that car, in design and function, but produced in 2022.

It's...well...*unimaginable*.

In the aviation industry, however, this is essentially what has happened. The bureaucratic sclerosis created by the FAA, with its serpentine certification process, has created stagnation within the industry and encouraged planes to stay the same. To anyone who understands the importance of innovation and progress, it is sometimes rather discouraging.

None of this is to suggest that there was anything mechanically wrong with the Twin Bee we flew in Winter Haven—merely that it was in fact an older piston–driven aircraft, which, by definition, made it more likely to experience problems any time it took to the air, particularly under the rigorous conditions of a simulated engine failure…at low altitude.

As we bounced noisily down the runway, I was reminded again of what an ungainly aircraft it was. It handled like your grandfather's Lincoln—sort of drifting and lumbering rather than rolling cleanly. It wasn't a tight or powerful plane; it wasn't nimble; it wasn't a *pilot's* plane. It was a big tail-dragging clunker modified for a specific purpose—and that purpose had nothing to do with flying smoothly or elegantly. It wasn't just a seaplane; it was a boat with wings. As we picked up speed, I wondered half-jokingly whether we'd even be able to get off the ground.

We weren't more than one hundred fifty to two hundred feet in the air, just seconds after takeoff, when Jim reached over and pulled back on the throttle, thereby initiating the simulated engine failure. This surprised me since I was familiar with protocol and figured we'd be going a little higher—as we had on the first flight with Tim. But again, I was also aware that the FAA guidelines are merely recommendations, not rules, and I'd been through enough challenging training exercises under less-than-ideal conditions to not let it bother me much. It was more like, *Okay, I guess this is going to be real-world training.* And anyway, there is supposed to be an element of surprise in a simulated engine failure because, after all, it's not something that ordinarily occurs. It's a malfunction that must be dealt with calmly and immediately. I looked out the window and saw trees and grass all around us, the sprawl of a housing development off in the distance. There was a lake to the left of the runway, and another lake a bit farther away. And I remember thinking in that moment that I would have to talk to Jim after we landed because I wasn't comfortable asking the rest of our guys to do the simulated engine failure at such a low altitude. I understood what he was trying to do, that he was an old-timer trying to test me in the most rigorous way possible, but there was an element of risk that seemed unwarranted. So, I made a mental note: for the rest of the day, five hundred feet, minimum.

In any conventional twin-engine aircraft with both propellors rotating in a clockwise direction, the left engine is considered the "critical engine." The descending blade of each propeller produces more thrust than the ascending blade. Therefore, in a conventional twin, the center of thrust is offset to the right of each engine. There is a greater distance, or arm, between the center of thrust and the longitudinal axis on the right engine than there is on the left engine; the greater the distance, the greater the leverage. Therefore, if the right engine fails, there is not as much yaw produced as there is if the left engine fails. Thus, the left engine is the critical engine; if the critical engine fails, the plane can quickly lose symmetrical thrust and fail to fly evenly, resulting in… well…a lot of bad things.

In a simulated engine failure, the instructor almost always "fails" the critical engine to ensure the most challenging scenario, so I figured that's what Jim had done. But the whole point of the exercise was to take nothing for granted—to work the problem methodically and clearly, step by step. No shortcuts, no rushing. In a survival scenario, the flight or fight instinct kicks in immediately, filling your body with adrenaline. And while adrenaline can save your life in a combat scenario, it can also interfere with rational thought. A pilot experiencing engine failure can't afford to panic. A perfect example of how to handle an emergency scenario is Captain Chesley Burnett "Sully" Sullenberger III, who miraculously landed a U.S. Airways plane carrying 155 passengers on the Hudson River following damage to both engines after it collided with a flock of birds upon takeoff. All passengers survived and only a few experienced any injuries at all. Luck was on Sully's side that day, but there is no denying that he and his copilot responded with extraordinary skill and composure in the face of overwhelming odds. And the movie version of their story did a terrific job of demonstrating how important it was for them to stay calm and follow protocol—right to the very end.

While I knew the basic layout of the plane and had studied the manuals for the Twin Bee, this aircraft was mostly new to me—after all, I had only spent an hour flying in it. It was not, as we like to say in aviation, an extension of the pilot. Nevertheless, I knew what to do when Jim pulled back on the throttle. The first step was to verify which engine had "failed." This is the most important step in a true engine failure because it's not unheard of for a pilot, in the throes of panic, to wrongly diagnose the problem and incorrectly shut down the good engine, thus creating a plane that is entirely without power. You have to take the time to check the engines, check the gauges, and make sure that you have correctly identified the failed engine before taking any further action. The urge to rush through the process is understandable but can be overcome through practice and patience. Special Forces training helped me realize the value of "slow and steady" as opposed to "fast and reckless," so I was relatively comfortable in this scenario.

But as I methodically worked through the process, something felt strange. I looked out the left side of the plane and saw that that propellor on the left engine had stopped moving.

"We have a failure in the left engine," I said.

Jim, thinking that I was merely playing along with the simulation, just nodded.

"No, Jim," I pointed out the window, "we have an actual failure on the left side."

Jim glanced out the window, said, "Oh, shit," and immediately took over the controls.

I still don't know what went wrong. Pulling back on the throttle should not have caused the engine to fail. Again, the automobile analogy is appropriate: if you take your foot off the gas, the car will decelerate and lose power, but the engine will not shut off. Unless, of course, there is something wrong with the engine. And yet, a months-long investigation that followed could not determine a cause. The FAA stated that there was no indication of a mechanical failure. But I was in the cockpit of that plane, and there was clearly a mechanical failure. The left engine stopped working, and the Twin Bee became almost instantly uncontrollable.

In most applications, an engine failure is not a death sentence. I had done many simulated engine failures in other types of aircraft, all without incident. I had experienced a real engine failure once previously, but that was mid-flight, which is a more manageable problem. An engine failure on departure, especially in this type of aircraft, is a worst-case scenario. Oddly, I felt almost serene as the reality of the situation sunk in. This I can attribute to military training, particularly in combat scenarios. The first time you're in a firefight, you respond instinctively—you try to stay alive by any means necessary. It's almost like tunnel vision takes over—you can only see what is right in front of you, your ears become plugged. But after you've been through it a few times, you begin to respond differently. You're able to utilize your training and think critically about the decisions you must make. It's like time slows down and allows you to process rather than panic. Moreover, if you are leading from the front in a combat scenario, it's imperative that you project a sense of calm—*everything is going to be okay*—because calmness is contagious. If you're a leader and you start screaming and freaking out, everyone else is going to freak out. The stress level rises and performance sinks.

So, while I certainly understood the seriousness of the problem when I saw that the propellor on the left engine was not moving, I did not overreact. And neither did Jim. Although he didn't say much, he remained calm as I began going through a checklist, verbalizing every possible solution to restart the engine or regain control of the plane using only one engine. Unfortunately, because the engine failure had occurred at such a low altitude, we only had a few seconds to come up with a solution. Jim said something about aiming for the lake that was not too far off in the distance, so that became our target—try to glide in, somewhat level, until we could reach water. Then maybe we could land on our pontoons. If not, the lake would be a more forgiving place to crash.

As we descended, Jim tried to maintain power—we couldn't have been going more than sixty knots, barely enough to remain airborne. The trees rose quickly in front of us, much more quickly than the lake. I knew from parachute training what this meant. We weren't going to make it. Anytime the foreground is rising faster than the designated landing zone, you're going to fall short of the target.

"We're not going to make it," I said.

There was another lake even closer, but it was off to our left, which meant that we would have to make a turn toward the direction of our failed engine. You might think this could work, since the right engine was still running, but the plane was already pulling hard to the left because the right engine was doing all the work. It was like trying to row a boat with only one oar in the water. Nevertheless, we began a slow banked turn to the left. Almost immediately, the left wing dipped, and the plane flipped over on itself and plummeted, nose down, tail up. A classic death roll. The next thing I saw were rooftop shingles approaching the windshield, faster and faster. We were diving straight into a residential neighborhood. Two thoughts went through my head. The first was a mental apology to my wife and kids for leaving them behind. And a goodbye.

I love you.

The second thought was, *I sure hope that house is empty.*

The assumption with any plane crash is that there will be some sort of explosion on impact, especially if the plane goes down upon departure with a full tank of fuel. It's almost guaranteed. But for some reason—pure luck, perhaps, or some quirk of physics—that didn't happen in this case. I opened my eyes seconds after impact, almost shocked to discover that I was still alive and not engulfed by flames. Somehow, I hadn't even lost consciousness. But I knew that it had been an intensely traumatic event, and figured I was probably bleeding internally or that I had been paralyzed. There was dark red fluid all over my face and hands, which initially I thought was my own blood, but when I couldn't find any gaping wounds, I figured it was probably hydraulic fluid that had spurted from the aircraft. I moved my hands. All good. I wiggled my toes. No problem. This did not mean I wasn't hurt; I was most likely in shock. I had been wounded in combat, so I was intimately acquainted with the body's ability to cope with trauma. On one occasion, several hours passed before I realized that I had been shot in the arm. Such experiences are not unusual on the battlefield or in any traumatic setting.

The front half of the plane's fuselage was mangled. But the boat hull of the airplane, designed to amphibious construction standards, had held its shape, which, in addition to the roof slowing our final descent, likely saved my life. I could barely see Jim to my right. He was neither moving nor communicative. Anticipating an explosion, I kicked at the hull of the plane—at the metal and the windows—until something gave way, and I was able to crawl out a hole and away from the plane. I found myself standing in the middle of what I realized was someone's bedroom. I could hear metal creaking. I heard people shouting. I also heard someone crying.

Before doing anything else, I pulled out my cellphone and called 911. I put the phone in my pocket, on speaker so I could have my hands free for first aid while working the scene and answering the dispatcher's questions. When it was clear that help was on the way, I went back to the cockpit to check on Jim. He was unresponsive. I checked his pulse. Nothing. Only much later would I dwell on the details of the crash and on the cosmic randomness of such events—how one of us could be killed on impact while the other walked away without major injuries. But there was not time for any of that now. I tried to shut down the fuel line to reduce the likelihood of an explosion. Then I followed the sounds of whimpering to a wall, where a teenaged girl was virtually embedded in the sheetrock. Somehow, the force of the crash had either catapulted her into the wall or caused the wall to collapse around her. She was alive but bleeding and barely conscious. I took off my shirt and used it to stanch the flow of blood from some of the wounds near her head and neck. By then, the house was beginning to fill with neighbors and family members eager to help.

"Someone find the circuit breaker and cut power to the house," I said, figuring this would at least reduce the chance of a fire. I had considered pulling the girl out of the sheetrock, mainly out of concern over an imminent explosion, but as minutes passed, that seemed less likely and was outweighed by the risk of moving her and exacerbating a possible spinal injury.

Within ten minutes, first responders had arrived on the scene. They extracted the girl from the wall and carefully placed her on a board and carried her to a waiting ambulance. There were other family members in the house, as well, but only the daughter was injured (she fully recovered, thank goodness). I walked outside to exactly the sort of scene you would expect when a plane crashes into the roof of a house in a suburban neighborhood: multiple ambulances, fire trucks, and cop cars. Dozens of neighbors standing around, gawking and talking. I walked out to the curb and sat down, suddenly exhausted and very sore all over. I was shirtless and covered with what I now realized was the congealed blood of my flight instructor.

I could overhear the cops talking to neighbors, gathering information, trying to figure out what had happened. It became apparent after a few minutes that no one had any idea who I was or why I was there. In an obviously tight-knit neighborhood of Black and Hispanic families, who was the random white guy sitting on the curb, and why had he been in the house, helping to stabilize the scene? And how the hell did he get there so fast?

"No, man, that guy was in the plane," I heard one of them say. I recognized him as the first neighbor to arrive. He had jumped right in to lend a hand.

"What do you mean?" the cop replied. "We have a report that the pilot is dead."

"No, not the pilot. He was the other guy."

"What other guy?"

"The other guy in the plane."

"What are you talking about?"

The neighbor threw up his hands. He was exasperated. "When I got to the house, that guy over there—that blonde guy—he was already in the house. I'm telling you, he was in the plane when it crashed."

"No shit?"

"No shit."

I had texted Tim as soon as I got off the 911 call to let him know that our plane had crashed but that I was okay. Little did I know that Tim had watched us take off and had witnessed the crash from the airport. We hadn't gone very far, after all, and the entire flight had lasted less than a minute. When they saw the plane go down, they immediately thought the worst: obvious death roll, nosedive into a house, no survivors. Tim was distraught and trying to process the scene as it unfolded when he saw his phone light up with texts from me. But he didn't look at them right away because he figured they had been sent prior to the crash. Only when he opened his phone and began dialing my brother to deliver the terrible news did he read the texts.

"Hey, Tim. We just crashed. I'm serious."

What I meant by this was, *Our plane just went down—I'm not kidding.* Dark humor is a big part of the military experience, especially in combat settings. You tell jokes to lighten the mood in sometimes brutal circumstances. It keeps everyone loose. I didn't want Tim to think that this was me playing some sort of practical joke, which is why I wrote "I'm serious." But Tim had witnessed the crash, so he knew I wasn't kidding, and what he thought I meant was, *I'm in serious condition.*

So, he commandeered an airport golf cart and took off with another guy from the airport, ripping through everything, driving toward the smoke from the crash. At this point, we were probably eight to ten minutes out from the crash, and I was sitting on the curb, just beyond the front yard, watching everything go down. Firefighters and EMTs were all over the place—they had the entire scene stabilized quickly. There was nothing more for me to do, so I just sat there and watched and waited and ran the whole incident over again in my head.

What the hell happened?

I noticed that there was a neighbor smoking a cigarette nearby.

"Can I bum one from you, buddy?"

I was shirtless, covered in blood from head to toe, with torn pants and only one shoe on.

Cigarette hanging from his lips, he looked at me like I was an alien and silently handed me a cigarette.

"Light?"

He gave me a light.

"Thank you."

We both looked back at the house together. He didn't ask me who I was or what had happened. And I didn't say anything. We just stood there in silence, watching the event unfold. I was pleased to see the young girl gently carried out of the house on a stretcher by four firefighters. She had injuries to her body and her eye that would require many months of recovery. Her family would eventually file a civil lawsuit against me, the airplane owner, and Jim Wagner as a result of the accident. I can't blame them—a plane did crash into their house, after all, and there isn't a day that goes by that I don't think of the accident and the trauma they experienced.

Suddenly, I saw Tim driving this golf cart from across the street through several yards, taking every shortcut possible. He jumped a hedgerow and popped back out onto the street near the crash and nearly rolled the golf cart as he pulled up to the curb. When he saw me sitting there—shirtless and bloodied, but very much alive—his eyes lit up and he smiled.

"Holy shit! You're okay?"

"Yup. Guess so. I mean, I'm here."

While I talked with Tim, another ambulance showed up and one of the EMTs began examining me and asking questions.

"What are your injuries?"

"Honestly, I feel fine. A little sore is all." I looked back at the house, with an obliterated airplane sticking out of the roof. It seemed impossible. "Based on the impact, I would think I've got some internal bleeding or broken bones...or something. But I just don't know."

"All right, well, let's get you checked out."

They put me on a stretcher and dragged me off to a nearby hospital for a bunch of tests and x-rays. They asked about the various objects that showed up on the scans—shrapnel, a bullet, and hardware from previous wounds and injuries—and they carefully went about the unpleasant task of cleaning me up, tweezing shards of glass and metal from my head and arms, sponging the congealed blood and brain matter from my hair to make sure there were no open wounds. After a few hours, a doctor came by to chat. He thumbed through my charts, looked at the scans, and shook his head.

"I can't believe I'm saying this, but if you want to leave, you're free to go."

"You didn't find anything?"

He shrugged. "Well, no. But you've been through a terrible crash and there could be something wrong internally that we just didn't pick up or that hasn't presented itself yet. I'll keep you here as long as you'd like to stay, as long you need in order to feel safe and comfortable. You can rest and recuperate, and we'll do some more tests just to be certain. That would be prudent. But medically, I don't have any reason to say that you *should* stay. And obviously, we can't force you to stay. So I am obligated to tell you that you can leave if you want. It's up to you."

"Okay, get me out of here."

In the ambulance on the way to the hospital, I had called Carmen to let her know that I was alive and well. I figured it wouldn't take long for her to hear about the crash and see pictures and news reports on the internet, and I wanted her to hear it from me first, so that she'd know that I hadn't been seriously injured. She was up in the mountains skiing with the kids at the time—like, literally out on the slopes—so didn't have great reception. But I was at least able to communicate the most important details: there had been an accident, but I was okay.

Carmen immediately went to the base lodge, where she'd have better cell reception, and googled the words "plane crash" and "Florida." As expected, she was treated to the sight of a Twin Bee's tail protruding from the roof of a house in Winter Haven, along with hastily written and appropriately horrifying accounts of the incident. Her first thought, appropriately enough, was that I had been lying and was probably badly injured but didn't want to worry her. We had both been through enough deployments to know the value of withholding otherwise frightening information. There were many things I saw and did in combat that I did not share with Carmen until much later…if then. Complete transparency is all well and good, unless it makes your partner's life even harder. On this occasion, I had told her exactly what I knew at the time, which was that I was alive and seemingly okay. A modicum of internet research seemed to indicate otherwise.

It wasn't long before Carmen was in contact with Tim and my brother, who was able to charter a private jet through a friend and fly to Florida from his home in Kansas City immediately. By the time I walked out of the emergency room, they were all there: Tim, Matt, the other guys from Bridger. On the way back to the hotel, we stopped at a Target to get me some clean clothes, and one of the guys bought me a T-shirt with a picture of a leprechaun and the words, "One Lucky Bastard."

An FAA official was already at the hotel when I returned, waiting to take my statement. I walked him through the entire incident and explained every detail as I saw it. Obviously, I did not know what caused the critical engine failure; four years later, as I write this, I still don't know. And neither does anyone else. I knew right away that there would be lawsuits—perhaps in multiple directions—and that is precisely what happened. Whenever there is a fatal aviation crash, lawyers tend to swoop in and begin cutting through indemnification clauses and legal precedence, hoping to hold someone—anyone—accountable. Fingers point in all directions—at the pilots, the airline, equipment manufacturers (it's not unusual for subcontractors who manufacture specific parts of a plane to be sued). It is endless and sad and often yields a settlement but no answers.

This is especially true in the case of a small accident such as ours. When a commercial flight carrying a couple hundred people goes down, the ensuing investigation is exhaustive, often stretching out over many months and even years. Given the magnitude of the loss and the impact it will have not just on the airline involved, but on the entire industry, this is understandable, even commendable. Families want answers and compensation. Potential customers want to know it won't happen again; they want to know they are safe. And as I've said, there is no mode of travel safer than commercial flying. The statistics repeatedly bear this out. Planes do fall from the sky, but they tend to be smaller aircraft that face less rigorous standards and upkeep, or aircraft whose pilots are less skilled or whose work is inherently dangerous. There are so many variables. But one thing is for sure: when a Twin Bee goes down in Winter Haven, Florida, killing only the instructor pilot, the investigation into that incident will be far less comprehensive than if it had been a sold-out Delta 737. Every day I think of Jim and his wife, Jinny, and their family. They have never gotten the answers they deserve and probably never will. They remain in my thoughts and prayers.

The FAA did not take long to determine that the accident was caused by "pilot error," which is always the result when a specific mechanical failure cannot be found. I don't know what that error could have been. We had been aloft for less than thirty seconds when the aircraft experienced a true critical engine failure. Is it possible that in pulling back on the throttle, the instructor caused the engine to stall? Well, anything is possible, but the fact remains: a plane in good working order should not lose power to either engine simply because the pilot pulls back on the throttle. If there was a mistake made, it was the initiation of the exercise at too low an altitude. That is not what caused the engine to fail, but it did leave us with almost no time to respond to what became an actual emergency.

I was reminded after the accident how fortunate I was to have such an amazing family and support network around me. It often takes these types of events to remind us of how lucky we are in this regard. Matt not only chartered a jet and made his way to Florida before I was even out of the ER, but he immediately jumped in to help with my legal defense against the many spurious lawsuits I would be—and still am—subjected to as a result of this terrible accident. My leaders at both companies, AVT and Bridger, Lee Dingman and Darren Wilkins, respectively, stepped in and ran things smoothly in my absence. My general counsel, James Muchmore, deftly helped to navigate our legal course as well. I was lucky in so many ways.

I was also reminded that it's often the routine times in life that are the most dangerous. One of the times I was wounded in Afghanistan, I was struck by a friendly ricochet bullet. I never reported it because I didn't want to get sent home and lose my team, and I didn't want the teammate who had fired the shot, a total stud who went on to a successful career as a SEAL, to be punished—officially or reputationally—by an accident that was in no way his fault. It wasn't even a tough or dangerous mission; it was a milk run, just like this training flight, but it went bad quickly. It's a good metaphor for life: it can go from a sunny day to a hurricane in a matter of seconds and turn your whole life upside down. Always be ready.

For me, the carry-forth from the accident was an intensified focus on safety and training. A few months after the accident, Viking lent us a Twin Otter—a far safer and more capable aircraft than the Twin Bee—for training and certification purposes. And everything went smoothly thereafter. For a time, I did experience a bit of PTSD, mostly in the form of concern about whether I was the human equivalent of a bad omen. I'd been through a number of close calls in combat—IEDs blowing up and killing or maiming someone close by but leaving me unharmed—and after this accident, I started to think, *Maybe it's not so good to be around me. Maybe I'm not such a lucky bastard, after all.* I worried about flying. I worried about driving if my kids were in the car with me.

I worried.

But after a time, the anxiety settled down and became merely fuel for vigilance. I became emboldened in my vision for the company. We would do good work, important work—and we would do it as safely as possible. No matter the cost.

At the time of the crash, Carmen and and I were the only two people who knew that she was pregnant with our fourth child, Walter (our third child, Bruce, had been born in 2017). It's been an emotional experience these last few years, realizing how close my children were to losing their father, and how close I was to leaving Walter in a position of never even meeting his father. Proximity to death can be a terrible and scarring experience. I don't wish it on anyone. Seeing friends, partners, compatriots dead or mangled—when just seconds earlier they were alive and well—is a gut-wrenching experience. It changes you forever. Yes, it can help to define the positivity and blessings in your life, but it can also destroy you. We see this in the epidemic of suicidal and homeless veterans. When your community and family embrace you and give you a solid emotional network within which to heal, you emerge from these traumatic experiences stronger and more resilient. You have a more complete perspective on the human experience. And, perhaps, you become a happier and more grateful person, determined to earn the sacrifice of those who died before you.

I have nearly been killed a number of times—IEDs, bullets, rockets, plane crashes—but it wasn't until I saw myself in Walter that I realized how many close calls I've really had, and how much I cherish the blessings we have as a family and as a nation.

CHAPTER 12

AIRTANKERS

"We have a saying here that I kind of like: 'mission first, safety always.' They go hand-in-hand. You can't safety yourself out of being able to accomplish your mission. You have to be able to accept some of the risk and mitigate that as best you can. Through professionalism and hard work. And education and knowledge."

—Bridger Chief Pilot Barrett Farrell

In the late winter of 2020, as we awaited the delivery of our first CL-415 Super Scooper, I could not have been more upbeat and excited about the future of our company and our ability to contribute to the mission of aerial firefighting. But aviation has a unique ability to smack you in the face with the reality of nearly crippling economic and administrative obstacles. I was only a few years into my second career, and while Bridger was a resounding success by any reasonable metric, there was still so much I had to learn—both as a firefighting pilot and a businessman.

One morning, as I read through our updated U.S. Forest Service contract, I saw a few items that looked like they might be problematic, but certainly nothing fatal. For the previous six years, we had been identifying and clearing hurdles through sheer force of will, and I figured this would be no different. Everything was generally as expected, and I felt we would meet the Interagency Airtanker Board's approval without much issue, and that they would allow us to fly our new scoopers anywhere for any agency. Smooth sailing, as they say.

I was wrong.

Deep in the contract, there was a brief reference to the fact that aircraft submitted for contractual work with the Forest Service must comply with MSG-3 standards. Specifically:

"Operator's aircraft must comply with MSG-3 as applicable."

"As applicable" being the operative phrase. A seemingly innocuous statement that we interpreted, as any reasonable person would, as being applicable to aircraft manufactured under the MSG-3 construct. But while this may not have been much of a concern to me, it was of great concern to others, notably our chief inspector, Tom Willis. Tom had come to Bridger with an impressive resume that included not just a career in the U.S. Air Force, but a deep background in certifying aircraft for Boeing and other major manufacturers. After carefully reviewing the contract, Tom and Steve Zinda, our senior vice president of maintenance and strategy, both expressed concerns and suggested we set up a meeting. They also requested that Darren Wilkins be present at this meeting. Ordinarily, it would have been extremely unlikely for all of us to be on-site in Bozeman at the same time, but this was in the early months of the COVID-19 pandemic when travel was largely halted. And of course, no scooper program meeting was able to be held without the presence of our director of maintenance for the Scooper program, Andrew Hill. Andrew is a force of nature—and not just because of his massive six-foot-two, 250-pound frame. He also finds ways to push through any bureaucratic obstacle to get things done.

"Morning, guys," I said as I walked into the conference room, unaware of the news that I was about to hear. "How are things progressing? Looks like we should have two-eighty-one here soon." (The tail number of our first scooper was 281.)

"Yes, hopefully in the next ten days," Darren said, his tone and demeanor somewhat less enthusiastic than I would have expected. "I'm still working with Paul Linse at the Forest Service to get us added to the Federal Scooper Call-When-Needed [CWN] contract this year. It is dragging on, but it sure doesn't help that we don't have a plane yet or a C of A [certificate of airworthiness] to show them. It's really making it hard for me to put the pressure on them when we don't even have an aircraft yet."

This made sense. Still, given the way dominoes had repeatedly fallen in the right order and place over the previous few years, I just presumed that this would work out as well. I remained upbeat, not because I didn't sympathize with Darren's concerns, but simply because I wasn't all that worried.

"Yeah, I understand," I said. "Along those lines, Tom and Steve have some concerns they'd like to discuss."

"Certainly," Darren said. "What's going on?"

"You remember the MSG-3 standard we spoke about a few weeks ago?" Tom began.

"Sure, Maintenance Steering Group-3 or whatever it's called? It's an airline standard that applies to airline production and maintenance standards." I replied.

Tom nodded. "That's right."

"Airline stuff," I said. "Doesn't apply to us because the scoopers are purpose-built."

This was 100 percent accurate: the CL-415 had been designed and constructed for the sole, or at least primary, purpose of fighting wildfires. And this aircraft was brand new. My assumption, then—and this was based more on common sense and a lack of institutional memory within our company—was that the Super Scooper would not be subjected to the same certification process required of an aircraft that had been designed for some other purpose. It made sense that a DC-10 converted from commercial passenger service to tanker service would undergo if not a more *stringent* certification process, certainly a different process.

"Right?" I said to a room that had fallen strangely silent.

Tom sighed deeply.

"Unfortunately, no. We talked with John Nelson at the Airworthiness branch, and he made it quite clear that the MSG-3 standard applies to our aircraft as well, even though they are purpose-built."

"Okay, so what does that mean?" I asked, fully expecting that one of the smart and capable guys in this room would provide a relatively quick fix based on the sort of unconventional thinking that that had helped Bridger overcome so many other challenges.

"It means we need to build a maintenance supervision program that conforms with airline standards," Andrew said, his tone flat and emotionless.

"Great, then let's do that," I said. "We don't have a lot of time, so let's get started right away."

I might even have clapped my hands together after saying this, like a coach drawing up the winning play on his clipboard in the final seconds of a basketball game.

No big deal. We got this!

There was a long period of silence as Tom and Steve looked at each other, mentally playing rock-paper-scissors to decide which one of them would inform the boss that he had no idea what he was talking about. Meanwhile, Andrew just stared at me blankly. Finally, Tom spoke up.

"Tim, this isn't something that can happen overnight. It can take years. We don't just write it up and turn it in. The MSG-3 standard is based on flight data, maintenance data, crash data from previous aircraft…and moreover, all that information has must be provided by the OEM [original equipment manufacturer]; it's not something we can do on our own."

Finally, it began to sink in. We were, in a word, fucked.

Shortly after this meeting, I started down the path of enlightenment—reading everything I could find regarding MSG-3, structural integrity, the philosophy and purpose behind these byzantine regulations, and, most importantly for the purposes of this book, the history of how these requirements came to be. The MSG-3 became my conduit to understanding the background of aerial firefighting, a lens through which I truly came to understand the roots of the industry—six years after I first became a part of it.

In the end, this standard applying to our aircraft wouldn't have made any difference for the fire season of 2020. Regardless of MSG-3, the FAA didn't get us a type certificate until July, which caused us to miss any hope of getting added to a federal contract for the season. With the entire country—government included—grappling with the impacts of COVID, remote work, and lockdowns, the factory was delayed, cross border shipments slowed to a trickle. The success of our program for 2020 depended on a nuanced and delicate dance between Transport Canada, Viking, Customs and Border Protection, the FAA, Bridger, the USFS Airworthiness Branch, the USFS Contracting Office, and the USFS pilot certification branch—a dance that would have been miraculous in a normal year. In an unprecedented year like this, it was simply impossible. Every single aspect of our business was snarled in delays, extensions, and unanswered phone calls.

In recent years, the USFS Airworthiness Branch has increased its contract requirements and airworthiness standards at a breakneck pace. This includes structural inspection programs, carding inspections, and process requirements—all good and necessary things. But they require time and coordination. Unfortunately, the relationship between industry and the USFS Airworthiness Branch is at an all-time low. After decades of shoddy aircraft and maintenance, officials at the USFS Air Worthiness Branch are understandably hyper-focused on contractor compliance. But in the new era of purpose-built planes and pressurized jet airliners, instituting onerous maintenance programs can be incredibly challenging from both a technical and legal standpoint. And it can take months, if not years, to work through the process. It's completely unrealistic to institute a new requirement in April and expect compliance by June. But that's what happens, and the consequence, for industry, is an inability to operate. Many companies have been run out of business by the airworthiness branch, resulting in animosity that seems to run in both directions. I truly believe that the vast majority of contractors agree with and support the initiatives and safety culture of the USFS Air Worthiness Branch. But they want to be part of the conversation and part of the solution so that these standards are crafted in a collaborative and realistic manner, rather than having the standards thrust upon them in a capricious and punishing way that ends up creating workarounds and adversarial positions between the providers and the customer. This is one of the many reasons the United Aerial Firefighters Association was formed in 2022—to encourage a more constructive relationship between industry and the USFS Air Worthiness Branch.

(A side note: In 2023, the AWB issued a directive to all U.S. helicopter operators that they had to comply with a U.S. Army flight manual requirement that conflicted with civil aviation standards. Furthermore, that requirement was considered confidential information by the Army and was not to be shared outside of military channels. Nonetheless, the AWB put this requirement into the contract and enforced its use, immediately putting the aircraft owners and operators into an impossible three-way standoff with their regulators (FAA), customers (USFS), and the United States Army. Not an enviable place to be.)

★ ★ ★

The entire concept behind having the MSG-3 standard inserted into USFS contracts was to prevent vendors from converting overused former military cargo or bomber aircraft that had been decommissioned by the Department of Defense into firefighters. Indeed, that was the very genesis of the aerial firefighting world in the United States: former military aircraft donated (or sold for pennies on the dollar) to agencies or companies who could then turn them into slurry bombers. Many of these notable aircraft included B-17s, PBY Catalinas, P-3 Orions, C-119 Boxcars, S-2 Trackers, and others.

The military orders extremely robust and capable aircraft. Once it takes possession of these planes, it flies them hard in austere conditions and maintains them to an operational field standard. However, it's important to remember that as professional and detailed as military aviation might be, it is a mission-oriented group, and that group will "make mission," regardless of the cost. Which means, especially if coming from combat duty or naval missions, these aircraft will almost certainly be "rode hard and put away wet," as we say. Meaning, in laymen's terms, they get the shit beat out of them.

The way it worked in the early days of aerial firefighting was that the military would decommission a set of aircraft and then either sell them at auction from the salvage yard to private owners or—if they were deemed still serviceable—conduct an inter-government transfer and "gift" them to an agency that could use the aircraft for firefighting. Both approaches have met with success as well as tragic failure.

The first path—private sales—normally happens in one of two ways. The aircraft are sold through a standard government auction process or the aircraft are mothballed and sent to Davis Monthan Air Force Base in Arizona, where they will sit in the desert for years or decades until they are no longer considered part of the nation's strategic reserve. At such time, private citizens can go visit the mothball yard and purchase old aircraft for, again, cents on the dollar. This is not an exaggeration. In this scenario, for a processing fee of approximately $4,000, anyone can buy a plane that the government bought for $30 million in 1970. It is, on the surface at least, an astonishingly good deal. But as anyone with even a modicum of experience in the aviation industry will tell you, buying a plane is the easy part. Getting it airworthy and flying it back to home base for eventual conversion into an air tanker is the hard and expensive part—or at least the beginning of the hard and expensive part. It can take years—decades, even—to get an airframe approved. Or never, as was the case with so many companies that tried to provide a new tanker solution. Many of them legitimately designed a bad system that was, for good reason, never allowed to take the field. Others were simply unable to survive the death by a thousand cuts that the bureaucratic process subjected them to.

Part of the reason for such an extraordinarily long and rigorous certification process stems from a legitimate desire to ensure the airworthiness of a plane and the safety of anyone who might operate it. But there also is an undeniable element of bureaucratic sclerosis, the result of decades of interagency dysfunction.

The FAA considers aerial firefighting as fitting into Part 91 of the agency's Code of Federal Regulations (CFR), which is essentially a category governing miscellaneous flight operations. In the FAA's view, aircraft utilized in aerial firefighting are public-use assets performing a public safety mission that is unique unto itself, and which the FAA does not have governing authority over beyond the requirements of CFR Part 91. However—and this is a very big "however"—the agencies that contract with us, including the Department of Interior, the U.S. Forest Service and CAL FIRE, insist that we operate under a myriad of commercially focused flight regimens from the FAA. These include Part 135, Part 145, Part 137, and others. This is akin to driving a car that you own to transport a load of medicine while under the employ of the local health department. The Department of Transportation (DOT) and the Federal Highway Administration (FHA) consider you to be a private vehicle, regardless of whether you are driving for hire, and treats you as such. But the local health department, wanting you to operate more professionally, tells you that you must operate as a commercial carrier under DOT standards, even though DOT does not consider you to be a commercial carrier since you are operating your own vehicle.

If just reading this makes you feel like you're drowning in alphabet soup, well, imagine how we felt in 2020 while waiting for our thirty-million-dollar scoopers to arrive as we wrestled with multiple agencies and sifted through thousands of pages of regulations and documents.

There is no shortage of anecdotal evidence to support the notion that governmental oversight, taken to the extreme, with a lack of coordination and communication between regulatory bodies, can impede progress and act as sludge in the gears of industry. But it's also true that regulation and oversight are necessary to ensure fair and safe businesses practices and to protect the citizenry from unscrupulous or negligent businesspeople, particularly in matters regarding public safety, such as aviation. It is also true that many of the logistical and regulatory issues we faced can be traced back to the Wild West days of aerial firefighting and its well-earned reputation as a rag-tag militia–oriented enterprise in which practitioners did the best they could with whatever tools they had at their disposal—where the envelope was routinely pushed to the tearing stage because, frankly, no one really noticed or cared.

Until, of course, they did.

This is not a criticism so much as an observation about an industry and a group of people I have come to love and respect. It is simply a matter of historical fact that aerial firefighting has never been on the cutting edge of aviation technology and probably never will be. The pioneers of aerial firefighting figured out how to fight fires without much help from anyone, armed only with the tools and technology they could create on their own. It was born as a baling wire, duct tape, and bubblegum kind of operation, and it wore that label proudly. The pioneers didn't have the finest aircraft or engineers at their disposal. What they had were planes left over from World War II and the conflicts in Korea and Vietnam—planes gifted to them by the military or sold at clearance prices. Even in the late sixties and seventies, we had not yet ushered in the era of foolproof commercial airline safety, so very few people noticed what was happening in aerial firefighting. They simply made do with what they had. The results were mostly successful…and occasionally disastrous.

It was, in many ways, a classic coming-of-age story, the kind you might find in any industry. Ingenuity, resourcefulness, and courage compensate for a lack of resources, leading to early and often heroic results. There is a period of impressive development and technology as everyone races to figure things out. For a while, you can keep pace and scale up. Then suddenly, you hit a wall, and what was viable yesterday is not viable tomorrow. In the case of aerial firefighting, that line of demarcation came in the 1990s, when the process of converting old military aircraft into tankers began to collapse under the weight of inadequate safety regulations and ethical lapses on the part of both regulators and business owners.

Frankly, with the benefit of twenty-twenty hindsight, it seems to have been a questionable conversion process in terms of engineering, safety, and oversight. Moreover, it suffered from being the very definition of an old boys' club, with a rotating door between the military, private industry, and government agencies like the Forest Service and CAL FIRE. To some extent, unfortunately, this still exists today, but it was certainly a much worse situation thirty years ago when planes began falling from the sky and aircraft that were supposedly meant to be used for firefighting wound up running drugs in South America. This is the sort of thing that happens when the interests of government and industry are aligned with each other and not necessarily aligned with what is in the best interests of the taxpayer.

It is, in short, what happens when the fox is watching the henhouse.

In the case of aerial firefighting in the eighties and nineties, and into the early 2000s, the things that should have mattered most—the safety of air crews and firefighters on the ground and the effectiveness of the mission—generally took a backseat to what was best for those on the sidelines: the bottom line of a company or the next step in the career of a government official, for example. The quickest, easiest, cheapest solution was too often the answer to a problem. And since the entire industry of aerial firefighting flew below the radar (so to speak), no one really cared or noticed until the environment became so toxic and dangerous that it became impossible to look the other way.

In the 1980s, the U.S. Forest Service instituted a program known as the Historical Aircraft Exchange Program. The agency, intent on bolstering an aging fleet that had recently grounded a number of C-119s that were deemed no longer suitable for service, acquired retired Air Force C-130A transport aircraft as well as retired U.S. Navy P-3 anti-submarine patrol aircraft. Contractually, these planes were supposed to be used as aerial firefighting tankers; however, in many cases, ownership of the aircraft was transferred illegally to private companies, which then either used the planes for other purposes or simply sold them at a profit since they were acquired by the Forest Service at a heavily discounted rate.

With good reason, the Historical Aircraft Exchange Program is commonly referred to simply as "The U.S. Forest Service airtanker scandal." It resulted in years of investigations and criminal and civil lawsuits, at the end of which two people involved in the scandal were sent to prison. In what amounted to a giant black eye for the Forest Service and the entire aerial firefighting industry, some of the acquired C-130s were found to have been used not for firefighting, but in covert operations conducted by the CIA.

Even more disturbing—and embarrassing—was the revelation that there were at least two drug trafficking incidents involving C-130s from T&G Aviation, one of the original contractors in the exchange program. T&G leased one plane to Trans Latin Air, which was indicted in 1994 as one of the aviation companies used by the Cali Cartel of Colombia. A second aircraft was seized by the Mexican government in 1997 after it was discovered transporting drugs.

By that time, the Historical Aircraft Exchange Program had long been deemed a failure worthy of abandonment. But its legacy lived on in the form of something even more disturbing than fraud or misappropriation: a series of horrific crashes involving air tankers. As often happens in the wake of a government scandal, purse strings were drawn tighter. The Forest Service embraced a thriftier model in which airtanker operators were selected through a process that inevitably rewarded the lowest bidder. This is fine if you're trying to cut costs and brag to taxpayers and elected officials that you can operate on a leaner budget. It is not so good for the companies that must then figure out how to make a profit after winning a contract with a ridiculously low bid, nor is it particularly good for the firefighters who must then cope with the fallout that inevitably occurs when the tanker companies stock their fleets with decrepit aircraft salvaged from junkyards or third-world countries and then cut corners on things like maintenance and repairs in order to reduce costs.

It is a brutal cycle of cost-cutting and shortcuts that can only end badly, which is precisely what happened.

On June 17, 2002, a Lockheed C-130A crashed after dropping retardant on the Cannon Fire near Walker, California. But it wasn't so much the fact of the crash as the spectacular way the aircraft went down. Coupled with the fact that the disaster was captured on video, it provoked attention. Records show that the plane departed Minden Air Attack Base in Nevada at 2:29 in the afternoon, loaded with three thousand gallons of retardant. It arrived at the fire sixteen minutes later and, per protocol, conducted a spotting pass in preparation for its drop. The plane then crossed a ridgeline and descended into a valley to conduct the first of two drops (each drop would release approximately fifteen hundred gallons of retardant). Video showed that after completing the first drop, the C-130's nose lifted as the plane ended its descent and began to pull away. At almost the same time, the plane's right wing inexplicably began to fold upward. Seconds later, the left wing did the same. Then both wings completely detached from the aircraft, leaving the fuselage to roll over and crash upside down near the fire zone. All three crewmembers were killed.

The investigation into the crash revealed several interesting and disturbing facts, not least of which was the plane's age: forty-five years old. The plane was part of the original C-130A production series, and it had been delivered to the U.S. Air Force in 1957 and was retired from military service thirty years later. In 1988, it was sold to the Hemet Valley Flying Service, where it was converted to airtanker capability. Hemet Valley then sold the plane to Hawkins & Powers. By the time it went down, the C-130 had accumulated more than 21,000 hours of airtime. The National Transportation Safety Board (NTSB) found evidence of "fatigue cracks" on the wings and determined the crash to have been the result of massive structural failure.

A little over one month later, on July 18, 2002, while the NTSB's investigation of the California crash was still underway, a second catastrophic airtanker crash occurred near Estes Park, Colorado. The aircraft, a Consolidated PB4Y-2 Privateer carrying two thousand gallons of retardant, was making a left turn in preparation for its eighth drop of the day on the Big Elk Fire when its left wing suddenly folded upward and separated from the fuselage. The plane became engulfed in flames and continued to roll leftward, eventually plummeting to the ground at a forty-five-degree angle and exploding on impact. Both crew members were killed. A subsequent investigation by the NTSB again found structural failure to be the cause of the accident. As with the C-130 crash in California, the PB4Y-2 showed signs of stress fatigue in its wing. The verdict was not surprising since it was an even older plane, built during World War II and used by the Coast Guard until it was converted to an air tanker in 1958. Like the C-130, the PB4Y-2 was most recently operated by Hawkins & Powers.

Those two crashes, combined with a similarly disturbing 1994 crash involving a C-130A that resulted in the deaths of seven firefighters, resulted in the Forest Service and the Bureau of Land Management combining to form an independent investigative "blue ribbon" panel charged with examining issues related to aerial firefighting in the United States. The panel's findings were sobering, revealing a pattern of neglect that left most of the industry's large airtanker fleet unworthy of flight. The fallout was necessary but deeply damaging to both the reputation and mission of aerial firefighting, with the National Interagency Fire Center temporarily grounding the entire California fleet. Of the forty-four large airtankers in that fleet at the time of the two crashes, only nine were able to remain in service. Two years after the crashes, the Forest Service terminated the contracts with all its large airtankers and tried to fill the resulting gaps with helicopters, smaller tankers, and military C-130s.

It is with good reason that the first ten years of the twenty-first century are often referred to as the "lost decade" in aerial firefighting in the U.S., as the industry struggled to overcome the stink of scandal and the pain of lost lives. I knew none of this when I moved to Bozeman and bought a plane with the hope of starting a small aerial surveillance company. Even a few years later, after Bridger had grown into a modest but not insignificant aerial firefighting company, I remained mostly unaware of the industry's checkered past, in particular as it related to the use of airtankers bought from the military or salvaged from various boneyards. But in the spring of 2020, as we found ourselves staring into the abyss of an almost incomprehensibly long and challenging certification process, I did all that I could to become educated on the subject. What I came to realize was that while aerial firefighting was a noble and deeply important mission, it was also a business that had proven itself to be in need of dismantling and rebuilding—from the ground up.

This is the sort of chaotic and devastating fallout that occurs when you create an environment of extreme cost compression. People look for other ways to make money because they don't think they can make a profit—or even stay afloat—flying on fires for the government. They'll rationalize unethical or even criminal behavior: *Hey, you guys have compressed prices so much that I have to buy these terrible planes and fly them for rock-bottom prices because I'm barely making it. How else can I make money? I guess I'll sell the parts on the on the black market...or I'll run drugs with the plane. Times are tough. I'll do what I have to do.*

A toxic environment emerged in the nineties, and while unscrupulous business owners were surely to blame for much of what occurred, various government agencies were also part of the equation. All that mattered to them was the bottom dollar: *If you crash, it's your problem, not ours.* Everyone looked the other way for as long as they could, until planes started breaking apart in midair, and suddenly, there were investigations and an awakening of sorts. Aerial firefighting needed better airplanes, better companies, and higher standards across the board. That kind of refresh was absolutely warranted and resulted in a far higher safety rating than the industry had ever known. It was a good thing and a necessary thing, but as with all transformative periods, it was far from a smooth process. And there were unforeseen consequences, some of which we encountered in the spring of 2020, as we prepared to put our new Super Scoopers into service.

There is a long history of knee-jerk reactions on the part of the federal government, leading to sprawling bureaucracies with extraordinary budgets and oversight power—whether you're talking about the birth of the Department of Defense at the end of World War II or the Department of Homeland Security in the wake of 9/11. Name your cataclysmic event and it is almost always followed by a massive influx of money and research and staff and regulations, most of which are well-intentioned, if not necessarily wisely implemented with consideration of future consequences.

For example, on February 12, 2009, Colgan Air Flight 3407, a regional flight carrying forty-nine passengers and crew from Newark, New Jersey, to Buffalo, New York, crashed in Erie County after entering an aerodynamic stall from which it could not recover. There has not been another fatal accident involving a U.S. commercial passenger airline since the Colgan crash. One could certainly argue that new and sweeping safety standards implemented by the NTSB following the Colgan incident contributed to this spotless record. But—and there is always a "but"—the new regulations, mostly surrounding pilot training and certification, did have some unintended consequences. For example, among the changes was a rule requiring all airline pilots to hold Airline Transport (ATP) certificates, which effectively raised the minimum required flight experience for first officers from 250 hours to 1,500 hours. Interestingly, while pilot error was considered a primary factor in the Colgan crash, both pilots had exceeded this standard (each having well over 2,000 hours) and held ATP certificates. In hindsight, it seems to have been an arbitrary solution to something that wasn't even a factor in the performance of either pilot, and the new standard soon led to an industry-wide pilot shortage that lasted for nearly a decade.

Similarly, the grounding of America's airtanker fleet following the crashes in 2002 was unquestionably warranted, and many of the policies rolled out in the ensuing years were needed: higher standards, better inspections, insistence on structural integrity—all that sort of stuff. But an unintended consequence was the creation of an extremely challenging, almost adversarial relationship between government employees charged with implementing and enforcing the new regulations and not just the vendors supplying equipment, but the first responders operating that equipment. It became us (the aerial firefighting industry) versus them (the government), which is never a healthy collaborative environment, particularly when tasked with such an enormously important and evolving mission as wildland firefighting.

Neither side was entirely right or wrong. There was enough illicit behavior and negligence to go around. But certainly, by the early 2000s, there was a sense that the agency folks viewed our entire industry as being rife with pirates and criminals in need of being kept on the shortest possible leash. It may not have been a universal feeling, but I think that attitude started to pervade the agencies and created a challenging environment for collaboration and partnership that persists today in many ways. And I know that's very frustrating for a lot of people in the industry who say, "Hey, we're on the same team, doing the same mission—why are you treating us like we're a bunch of crooks?" And the agencies' answer is, of course, "Well, I remember when you *were* a bunch of crooks!"

In addition to meeting rigorous standards required for structural modifications of the airframe, the bombing system must pass the "grid test," which is literally a long line of plastic dixie cups placed along a runway in Arizona. The tanker being evaluated must "pass the grid," meaning that when it drops its load at a specified coverage level, each cup must be filled with a certain amount of retardant.

As the tanker drops its retardant on the grid, it is graded at each coverage level ranging from one to eight—one being the least amount dropped, eight being the most. In an almost symbolic and laughable nod to the archaic nature of so many standards in the air tanker world, the coverage level grading mechanism derives from the bombing density of a B-24 Liberator bomber in World War II. Coverage level eight means all bombs in the bomb bay are released at once out of the aircraft for maximum bomb density on target. Coverage level one means only two bombs are released at a time to create a linear bombing pattern with less density.

Under the best of circumstances, the process for air tanker certification is daunting and time consuming, requiring inspection and approval from multiple regulatory organizations—the U.S. Forest Service, FAA, Bureau of Land Management, Interagency Airtanker Board—that do not necessarily agree on the letter or intent of the law, which can complicate their guidance and expectations. The fact that we were seeking approval of a new airframe and importing these aircraft from a Canadian manufacturer complicated matters further. On top of that, of course, was the sudden emergence of COVID-19 and a nearly worldwide shutdown in the spring of 2020 just as we were preparing, or at least hoping, to put the CL-415 into service. You couldn't have picked a worse time to try to complete the international transport of an aircraft going to multiple different federal agencies to authorize its operation. And, of course, wildfire season waits for no one. April, which is when this whole process began, is "go time" for us, and we were eager to wage battle with the new scoopers at our disposal. Even after we were made aware of the MSG-3 standard and immediately began coordinating with various agencies to provide them with everything they needed to expedite the certification process, we felt the timeline was doable—aggressive, but doable.

COVID changed everything.

In the early days of the pandemic, a large percentage of government employees immediately transitioned to remote work. I'm not tossing criticism at anyone for the way they handled a once-in-a-century pandemic—everyone was doing the best they could under extraordinary and unforeseen circumstances—but the shutdown occurred at precisely the time our company was most dependent on interdepartmental and interagency cooperation. We had to somehow get all these different offices to come together and examine our application, at the center of which was a unique aircraft that had never been certified in the U.S.—and it had to be done at warp speed while everyone was working from home and the world was melting down.

Looking back, it's hard to say exactly where the lines of communication became disrupted. We had not stumbled blindly into this process.

You don't spend tens of millions of dollars on new aircraft unless you are confident that they will pass certification and be put into service in a reasonable time frame. Moreover, you don't convince investors to back your purchase without those assurances. We knew that the CL-415EAF model had never been certified and therefore the FAA would have to amend the older type certificate on the airframe to guide the certification process. But we had been conversing with the FAA and other agencies for two years and had been led to believe that while the guidelines were being written for the scoopers, we would be able to temporarily operate on an older but still applicable certification. So we took a huge amount of risk based on verbal communication with the FAA and Forest Service. And what I came to realize is that what government folks will tell you in conversation does not always translate to what they will do in writing. Things get bogged down, and that's how you end up on the phone, pleading your case to senators and congressmen, some of whom are helpful and some of whom tell you they will try to address your problem sometime in the next year or two.

In the end, it all worked out. Sort of. We got the planes imported and certified by the FAA and put into service by the middle of the summer of 2020. That was a great day for all of us at Bridger. But it was also a bit like bringing in a star quarterback in November: it's a little too late. You build your team in the offseason so that you're ready to go on opening day. And the CL-415EAF was supposed to be our star quarterback. Moreover, we didn't receive Forest Service certification until the following year—when we had to go through the entire process again—so in the summer of 2020, we worked exclusively on state contracts.

But the consolation prize—and it wasn't a small one—was the scooper's stellar performance in a range of applications. One more tool in the toolkit; one more weapon in the ever-expanding fight against wildfire. And, for me, significantly more knowledge about how the business works and how it interacts with government, and, hopefully, how we can all do our jobs more efficiently.

At the end of the day, we're all on the same team.

CHAPTER 13

THE INDUSTRIAL REVOLUTION

In the wake of the "lost decade," many of the reforms mandated by the blue-ribbon panel were eventually adopted, regulations were changed, companies went under, and the face of the industry evolved rapidly. What emerged from the smoke plume was a more heavily regulated and professionalized industry that adopted many of the operational safety standards of the airlines and the military to ensure safer operations within the aerial firefighting community. But it's important to remember that firefighting is *not* the airlines or the military. Our mission exists in a paradigm that actively subverts prior planning and negates many of the processes and procedures that make safe operations in other aviation paradigms possible.

To this day, tension exists over whether the regulations and standards impede the rapid and agile response to new fires. What we see in 2023 is a backlash by many state governments against the heavy hand of regulation that federal agencies have placed on aerial firefighting. Many state departments of natural resources (DNRs) and forestry groups have grown frustrated by what they feel is an over-centralized bureaucratic nightmare that stands in the way of getting the air support they need when their state is burning. Historically, states would fight fires with ground resources but would rely on the feds for air support. With the decrease in fleet size and a challenging dispatch process that slows the launching of assets when requested, many states are now building their own firefighting fleets with direct contracts in a way that is changing the dynamics of our small industry.

In the 2000s, as companies and agency officials reoriented their methods around a more structured firefighting industry, one thing became obvious: the U.S. was not going to have as many air tankers at its disposal. Going from nearly fifty large air tankers in the late 1990s to low teens around 2010, the U.S. experienced a new reality in resource scarcity. Requiring airline-compliant aircraft was going to move the cost of doing business into an entirely new league. We aren't talking about a 20 or 30 percent increase; we are talking about a 100 percent increase in cost—or more. Much more. This would, of course, remove many of the historical players from the field practically overnight.

Operations went from using retired B-17s, P2V Neptunes, Boxcars, and others to buying and modifying commercial grade jet airliners like the DC-10, MD-87, Bae146, 737, RJ-85, 747, and others—all pressurized multiengine turbojet aircraft with complicated parts supply, maintenance support procedures, and highly pedicured staffing requirements. This was no longer barnstorming ag flying or anything close to it; this was a new level of capability, and it ushered in a new era of aerial firefighting. This era was rooted in the U.S. Forest Service's Next Generation Air Tanker program, which encapsulated at a high level all the requirements for companies to now operate airline-quality jet powered aircraft and the affiliated staff, facilities, and tooling.

While aerial firefighting in the U.S. proceeded in a decentralized manner beginning in the 1950s, our Canadian partners to the north were evolving their aerial firefighting capacity in a more industrial manner. The Canadian approach was largely a government enterprise from Newfoundland all the way to Saskatchewan. These provinces had large fleets of Super Scoopers with a squadron designed to immediately attack fires with water. But in British Columbia, a different model was developing, and it looked a lot like the United States' model with a bit more centralized government control. Core to this was a firefighting legend named Barry Marsden, a consummate entrepreneur who started Conair with a few converted WWII dive bombers and slowly developed it into the largest aerial firefighting company in the world, routinely setting the standard for industry-wide operations. Conair developed around the timber industry of British Columbia, where millions upon millions of acres of dense forest land held trillions in economic value for its owners and the people of the region. To let it burn was simply not an option. And although Bridger Aerospace theoretically competes against Conair every day in the business arena, over a fire, we are one team. I deeply respect the quality and capability of the company Barry built and that Matt Bradley, a former F-18 pilot, now leads. I'm not sure they share the same opinion of me (being a young upstart in their old sandbox) but for all of us, it's a privilege to be in this business.

Even as aerial firefighting companies were forced to evolve, adapt, and become more professional, many of the strong personalities that formed the early days of firefighting remained. Men like Barry Marsden, Wayne Coulson, and Jack Erickson still drove the show in a very real way, and they weren't going to be tamed by new regulations. In one encounter, as Erickson Aero Tanker was taking its newly modified MD-87 through testing, concerns arose over retardant ingesting into the plane's rear fuselage-mounted engines of the MD-87.

Jack Erickson is a legend in the aviation world. His exploits in developing the Erickson Air Crane—adapting it to all manner of extreme use cases, from placing air conditioning units on top of skyscrapers to building custom snorkel tank technology for airborne water tanks—are well known. Hailing from a timber family in Oregon, he had a vested family interest in protecting his family's resources. Already using helicopters for logging, he built one of the largest and most respected helicopter companies in the world, Erickson Air Crane, from one helicopter into a massive global rotary wing operator. With the upgrades required for the new air tanker contracts, there was an opening for Erickson to move into large fixed-wing operations as well. Erickson Aero Tanker's MD-87 was born from this.

But as the MD-87 went through testing, the evaluators were concerned that as the swept-wing aircraft was dispensing retardant, the laminar flow coming off the rear part of the wings' wash pulled the retardant up and into the intakes of the two tail-mounted Pratt and Whitney JT8D engines. This could, potentially, cause major problems, not least of which was corrosion to the innards of the low-tolerance engineered parts of a multistage jet turbine engine. A serious side effect could be a flameout of one or both engines if enough retardant was ingested and smothered the compressor stage. A jet engine goes through almost unimaginable testing to prove its reliability for operations. It will sit on an engine stand in a wind tunnel and run for hours or even days at a time. Additionally, environmental testing will include spraying literal tons of water, ice, and snow into the engines to ensure they will run in all conceivable combinations of weather conditions and temperatures. At times, over eight hundred gallons of water per minute will be injected directly into the engine intake to test its mettle. It's quite a remarkable process that has produced one of the most dependable and reliable mechanical inventions in the history of the world: the jet engine. The manufacturers of jet engines today consist of a small group of large industrial conglomerates that include General Electric, Pratt and Whitney (United Company), Safran, Rolls Royce, and Honeywell.

However, one thing the manufacturers did not spray into their engines during testing was phosphate-based fire retardant. Being a heavy sulfate-based chemical mixture, it will not simply be evaporated by the heat in the engine. That is exactly the point of fire retardant—to put out fires, not be evaporated by them. So, a major focus of the new Next Gen Air Tanker inspection criteria was the retardant ingestion potential of jet engines. In the case of the MD-87, something in the testing process concerned inspectors, and they said as much.

What happened next was classic American aerial firefighting lore. After they filed their report on the MD-87, the small team of government inspectors was promptly informed that the MD-87 would be cleared for duty. After some confusing back and forth, they were told in no uncertain terms that the aircraft would be certified for use and that was that. Confused as to why this agency-adopted process was now, apparently, being circumvented by their superiors, the inspectors went about their business and scheduled the contract kickoff meeting with the Erickson team. All of this is accumulated by the rumor mill in the aerial fire community, so the details are hard to confirm. But as they entered the meeting room in Oregon, the inspectors were surprised to find just one attendee in the room waiting for them: Jack Erickson himself.

"Good morning, Mr. Erickson, are we having anyone else join us today?" one of the inspectors asked.

Jack looked across all of them.

"No, we are not," he said calmly. "And furthermore, if any of you ever fuck with me again, I'll have your asses handed to you on a plate. I built this industry from scratch, and I know what my planes can do. Don't think you can come in here and tell me any different. Now get out of my office and stay out of my way from now on."

Upon phoning their superiors for guidance, the USFS Inspectors were simply told to go about their business.

How much of this popular industry tale is true is hard to verify. What is true, however, is that the MD-87 is a robust Douglas airframe that has been successfully fighting fire for decades. It has its detractors, as all platforms do, but it has an undeniably solid performance record.

★ ★ ★

The C-130 has been the subject of more fatal crashes per capita than any other large fixed-wing firefighting aircraft. Yes, many C-130s have entered fire service, and, yes, many of them come from hard-flown military backgrounds, leaving them structurally compromised. The C-130 is a legend of aviation, an absolute workhorse of the American and Allied military universe. I have jumped out of many, static and freefall. I have even parachuted out the back of a C-130 with a boat! I love the aircraft and its capabilities, but its firefighting safety record has been concerning and is the source of many of the modernized regulations that restrict former military aircraft. The only remaining private operator of C-130s for firefighting in North America is Coulson Aviation out of British Columbia. One of the earliest and most innovative companies in aerial firefighting, the Coulson family owns and operates the business to this day with their flagship, the C-130 Air Tanker.

The Coulson family operated the largest seaplane in the world for many decades, the Martin Mars, sister ship to the Hawaii Mars. These aircraft were gargantuan in their stature, and to this day, they reign as the largest seaplanes ever consistently operated (one larger aircraft was built in France but never operated actively, and there have been other experiments, such as the Spruce Goose, that were never sustained). The Mars aircraft were designed and built by Howard Hughes in the years following World War II. A consortium of forestry companies ended up acquiring all four planes after they fell out of use, and they were converted into firebombers in the 1960s. The Coulson team, like Jack Erickson, found its firefighting roots through its family's involvement in the timber industry along the west coast of the Americas.

The Mars aircraft were a wonder of engineering with a 200-foot wingspan, 90,000-pound gross weight, and a payload of 150 personnel—it was an amphibious beast. Designed and built to enhance our logistical footprint in the Pacific during World War II, the production of the Mars ended up missing the war, so it had a brief service life in the Navy (1944–1956), then moved into firefighting. With four massive 2,500 horsepower radial piston engines, the joke about the Mars was that you filled the oil and checked the fuel. It had walkways inside the wings so that crew members could check the oil in the engines mid-flight during its fourteen-hour legs across the pacific.

The Martin Mars would end up being the source of controversy later in its life, but at the time of its build, it was an amazing feat of engineering. The Coulson family have been consummate innovators in the aerial firefighting space, developing the 737 Air Tanker (Dubbed the "Fireliner"), the night helicopter quick reaction force capability in

★ TIM SHEEHY ★

Southern California, and the continued development and enhancement of their C-130 air tanker program. Like any successful business, they have their detractors and enemies, many of whom were more vocal in the wake of their tragic C-130 crash in Australia in 2021, which killed the entire crew of three. The subsequent investigation of this crash concluded that pilot error was the cause and blamed "spatial disorientation" as the reason. As was the case in my crash, blaming the dead pilots is usually the easiest way out for all involved, with the exception of the families who will spend an eternity wondering what happened in the final seconds of their husbands', sons', brothers', and fathers' lives. This crash was further scrutinized three years later when a second Coulson large air tanker crashed in February 2023, also in Australia. Thankfully, both pilots walked away from that accident.

★ ★ ★

In the wake of the air tanker scandals of the late '90s and the structural failures leading to the crashes in the early 2000s, the U.S. Forest Service undertook the Next Generation Air Tanker program. This program, controversial at the time, would end up being one of the most impactful pivots in the history of aerial firefighting. The agency was boldly making the decision to ground over forty of its large air tankers, almost all surplus military aircraft, and nearly the entire large fixed-wing fleet in order to mandate the development and implementation of a very strict structural life inspection program. The stated purpose of this was to ensure that the already risky job of aerial firefighting was not made more so by the usage of salvaged military aircraft with questionable airworthiness. Embarking on the Next Generation Air Tanker program effectively spelled the end of the Wild West era of American firefighting and ushered in a new era in which fewer, larger, more well-funded companies would take the place of some of the barnstorming outfits that pioneered the profession.

★   MUDSLINGERS   ★

It's important not to place "blame" on these early pioneers for creating the best capability they could given the budgets and parameters they were allowed by the governing agencies. It's also important to match the trend line of professionalism and safety of fire aviation with that of the broader commercial aviation market as a whole. Remember our discussion of commercial airline safety? Well, it was only about a decade later that aerial firefighting began adopting many of the same policies, procedures, and standards that improved safety and reliability in that industry.

It may sound like a small transition, but the significance of this next generation push cannot be overstated. Operating a handful of retired military warbirds under a relatively hands-off inspection process was characterized by the owner-pilot-mechanic construct. The planes were typically large, unpressurized, piston-powered military transports. It wasn't unusual to see a grease-covered mechanic banging on one of these aircraft with a hammer to get it running again! Transitioning to the type of aircraft specified in the next gen contract meant a leap forward to a new era typified by multi-engine, jet-powered, pressurized, commercial-sized aircraft like those seen in commercial passenger aviation. In fact, it was just the type of safety margin improvement in the airlines that drove the USFS to their adoption of airline quality aircraft. The MSG-3 standard of airworthiness was an exceptionally stringent process that governed the design, engineering, development, production, maintenance, and ongoing inspection of aircraft produced under that construct. This leap required a company to cover seventy years of aviation progress—from World War II-era aircraft to twenty-first century jet age—in a matter of a few years. The financial implications of that were huge; from training to staffing to facilities to aircraft, it was a massive leap that many companies simply couldn't make.

The next gen shift, like all evolutionary transitions, was necessary, but it invariably created casualties along the way. Many of the early pioneers of aerial firefighting in the U.S. couldn't afford the transition to the modern age. Companies like Black Hills Aviation, Hawkins &

Powers, and Minden Air Corp, among others, either went under or were acquired by companies with deeper pockets.

Neptune Aviation acquired Black Hills, leading the transition out of the venerable P2V Neptune (a 1950s-era sub hunter) and into the flagship next gen air tanker, the BAe 146. Neptune has continued to grow that fleet and has been one of the stalwart large air tanker operators that keep America's large fixed-wing capacity at a viable level. Based out of Missoula, Montana, they have an impeccable operational record and are the poster child for the successful transition from the old ways to the new ways. The development of the BAe 146 into a tanker, however, did have controversy around it—as almost everything does in aerial firefighting. The original BAe 146 design is said to have been pioneered by Len Parker of Minden Air Tanker out of Minden, Nevada (coincidentally our first basing location with our scoopers). Len reportedly made several unapproved modifications to his BAe 146 that ended up making his aircraft unairworthy in the eyes of the FAA and thus the Forest Service, ultimately killing his company, but opening the path for a more established operator, like Neptune, to carry the torch forward with the BAe 146.

Len's is the classic story of the aerial firefighting barnstormer. A jungle fighter from Vietnam, he emerged from the war troubled, as many veterans do. But, like me, he found a mission in aerial firefighting. Working with other Vietnam veterans, he took advantage of the military aircraft transition program and slowly built Minden Air Corp into a respected regional firefighting outfit. Unfortunately, Minden become yet another example of a company that couldn't make the transition from the barnstorming days of the '80s to the industrial days of the 2000s. They went under after buying a small fleet of BAe 146s and trying to convert them to airtankers. Despite congressional support and local champions, the red tape of the USFS Airworthiness Branch and the FAA proved too formidable a foe. Today, Minden Air Corp and Len are largely selling off their parts piece by piece.

Meanwhile, the new generation of air tanker providers, performing on a level commensurate with commercial air travel and military operations, is a mixture of reimagined old-school companies like Neptune, Air Spray, and Erickson Aero Tanker, industry stalwarts like Coulson and Conair, and newcomers to the field like 10 Tanker and Bridger Aerospace. These newer providers of tanker capacity are widely regarded as being a vast improvement to the legacy model of air tanker operations that characterized the twentieth century. Like the rest of aviation, the safety margins, operational excellence, and effectiveness of today's air tankers far exceed what has traditionally been available.

Recognizing this new reality, it is more important than ever to give credit to the pioneers who created the capability that many now take for granted. Those men built amazing aircraft that gave the world a lifesaving capability. That kind of progress never happens without risk, failure, and controversy. In fairness to those early pioneers, it should be noted that they played a game in which the rules were set by the customer and the regulators. Price became the primary concern to contracting agencies in the '70s and '80s. Like a living organism, industry will respond to natural market forces. When the consumer of a product prioritizes a certain set of needs—in this case, price per day and hour to contract an aircraft—they will drive the behavior of that industry. Unfortunately, things rarely change until a catastrophe requires it. The crashes in the early 2000s forced that change and ushered in a newer, safer, more productive era of aerial firefighting.

★ ★ ★

While tankers and, more recently, scoopers tend to get the lion's share of publicity in the aerial firefighting world (and indeed, these resources have been the focus of this book), we would be remiss in not acknowledging the significant and consistent contribution of helicopters throughout the years. On almost any large fire, you are likely to see a mix of tankers, scoopers, and helicopters sharing air space, all under the direction of air attack. Like every asset in wildland firefighting, helicopters have their place and are arguably the most versatile and critical class of aircraft in wildland firefighting. From moving elite hotshot crews and other firefighters, to aerial ignition of controlled burns and bucket drops, helicopters are the connective tissue of the wildland firefighting community. For example, in parts of Southern California where access to large bodies of water might be limited, a helicopter could be more useful than a scooper...or a tanker.

"That area is spectacular for helicopters because there are thousands of small cattle ponds and golf courses, even swimming pools," said Jason Robinson of AeroFlite. "California is perfect for helicopters, and they've been expanding light operations there with Coulson's quick reaction force."

The quick reaction force (QRF) program, born in 2021, is a joint venture between the Los Angeles County Fire Department, the Orange County Fire Authority, Ventura County Fire Department, Southern California Edison, and Coulson Aviation. As of this writing, the QRF employs four of the industry's most versatile firefighting helicopters: two CH-47 helitankers that can each carry up to three thousand gallons of water or retardant, a Sikorsky S-61 with a one thousand gallon capacity, and a Sikorsky S-76 intelligence and recon helicopter, along with a mobile retardant base that can actively mix up to eighteen thousand gallons of fire retardant per hour. All four of the helicopters have night-flying capabilities.

"The combination of night-flying capabilities—night hovering, nighttime retardant filling, and dropping—is what is remarkable and unique," Troy Whitman of SCE, a twenty-nine-year veteran of fire management, said in 2022. "Taking the full extent of aerial firefighting and going nocturnal is unprecedented anywhere in the world."

Even when they do not carry retardant or water, helicopters can be vital to the cause in ways that no other firefighting aircraft can match. And that is in the realm of search and rescue.

Consider, for example, the California National Guard helicopter crews that rescued more than two hundred people from a fast-approaching forest fire in the Sierra National Forest in September 2020—a mission that the pilots involved labeled the most dangerous of their careers.

The crews were summoned to duty when the fast-moving Creek Fire engulfed the area around Mammoth Lake Reservoir, northeast of Fresno, stranding some two hundred lakeside campers whose escape routes suddenly had been rendered impassable. Utilizing a CH-47 Chinook and a UH-60 Blackhawk, the guardsmen, with night vision technology, flew through heavy turbulence and smoke and fire so thick that it made even some of the most seasoned crew members nauseous. They settled less than one hundred feet from the fire and extracted roughly sixty-five of the campers, some of whom were suffering from burns and smoke inhalation. As the fire raged, the helicopter crews made two more flights each, eventually rescuing all the campers safely.

Kipp Goding, who piloted the Blackhawk, described the mission as the most dangerous he had flown in his twenty-five years as a pilot—including combat missions in Iraq and Afghanistan.

"The stress and the added workload of going in and out of that fire every time is definitely by far the toughest flying that I've ever done," Goding told ABC news. "Every piece of vegetation around that lake, as far as you could see, was on fire."

★ ★ ★

As noted, there is a long and sometimes complicated history between the military and aerial firefighting, with overlap and influence on everything from piloting to equipment to tactics. But it's worth mentioning that many of the most capable and stalwart aerial firefighting aircraft have been former Navy aircraft. With the demands of carrier operations and the challenging weather of a maritime environment, naval aviation requires an additional level of structural integrity than does a typical military airframe. If you have ever seen a video of a carrier landing, then you will understand why this is true. The successful landing of an airplane on an aircraft carrier requires a level of pilot precision that exceeds almost all other areas and requires structural integrity on the airframe that far exceeds that of typical runway operations. For example, if you tried to land an F-16, the U.S. Air Force's mainstay fighter, on the USS *Nimitz*, the plane would be seriously damaged upon landing and would potentially be destroyed. A simple Google search of an F-16's landing gear versus that of a carrier-capable F-18's landing gear will show you the difference in the aircrafts' heft.

But landing gear isn't the only part that requires reinforcement. Every facet of the airframe—especially areas with critical structural integrity, such as wing spars, fuselage attachment points, and payload bays—has significantly higher impact and stress tolerance than a normal airframe. Of course, what this creates is a heavier, more robust plane that may not be as nimble or fast as the Air Force airframe, but it does have extraordinary reliability.

Even independent of carrier ops, naval conditions require robust construction. For example, the PBY Catalina, the seaplane workhorse of the U.S. Pacific theater in WWII, could withstand seventy-knot airspeeds across rough seas. If you have ever ridden a speed boat and felt the shudder when the hull hits the broad side of a wave head on, you know what I'm talking about. Now increase the speed, the weight, and the size of the vessel, and you can imagine how taxing it is on an airframe to be slammed repeatedly like that in active seaplane operations. It requires tremendous structural integrity.

Look at this way: if the Air Force builds a Camaro, the Navy builds an F150. And that's why, from the earliest days of air tanker operations, former naval aircraft have had a storied place in firefighting fleets around the world. And that's why the scooper has been an answer for so many. Naval aircraft structural construction quality has been the standard for a lot of seaplanes. The requirements for the fuselage to withstand the rigors of waves slamming into them at over a hundred miles an hour has become a defining quality in naval aircraft, firefighting aircraft, and almost all amphibious aircraft. And when the plane that I was flying went down in Winter Haven, Florida, this structural heft would end up playing a far more important role in my life than I could ever have imagined. After my crash in the Twin Bee in 2019, many folks said the strength of the sea plane fuselage saved my life. Much like getting in a cast iron tub during an earthquake or tornado is recommended, crashing in one also has its benefits.

CHAPTER 14

WITHDRAWAL

The 2021 fire season was the busiest and most destructive in U.S. history, with more than 48,000 wildfires consuming 6.5 million acres according to the Department of the Interior. Fifteen wildland firefighters lost their lives battling fires; thirty-three civilians were killed. Nearly five thousand structures were decimated. It was the longest, most exhausting season in the annals of American wildland firefighting, and the first, for me, that felt less like a season than a ceaseless, round-the-clock siege.

For those of us at Bridger, the season was also complicated physically, emotionally, and logistically by events occurring simultaneously half a world away—events over which we had absolutely zero control. Bridger is a company with many veterans of the U.S. armed services among its employees, including our founding members, so it was with no small amount of interest that we all watched the events in Afghanistan unfold in the summer of '21. I've already stated my feelings on the origins of the Afghanistan conflict, so there's no need to get into a philosophical diatribe here. But the ending, I think, is fair game for assessment.

We had known for several months that a withdrawal of American forces from Afghanistan was inevitable. The Trump administration had established a May 2021 deadline, and the Biden administration had vowed to adhere roughly to that deadline. Politics aside, it seemed simply like bad military and foreign strategy to pull up stakes after twenty years and withdraw completely from a country where thousands of American soldiers had given their lives to stabilize the country. Whether the mission had been noble or mostly folly was beside the point now. We were deeply invested, and protecting that investment was neither particularly complicated nor difficult. Even after the official pullout started to materialize in July, most veterans (myself included) thought that the administration would embrace a late-game shift in strategy—that political posturing would give way to common sense. Everybody knew we weren't ready to just walk away cold turkey. Even higher-ranking officials said, "No, it's not going to happen."

We all figured that we'd do what we always do: we'd get to the one-yard-line and realize things are not quite as stable as we'd like, cooler heads would prevail, and someone would make a decision to keep a small but visible number of troops—maybe a thousand people—in the country to help maintain order and, frankly, prevent the mass slaughter of civilians or opposition forces, some of whom had been friendly to the U.S. for two decades. There was no shortage of precedent. Desert Storm, Bosnia, Panama, Korea. With the notable exception of Vietnam, we've kept active-duty troops on the ground in every country with which we have been militarily engaged for more than a century. We had tried a complete withdrawal from Iraq in 2011, and stabilization quickly gave way to chaos. ISIS started burning people alive and beheading hostages on camera, so we invaded a third time, fought some more, and achieved stability again. This was no small achievement, and, one would think, a lesson in both humility and logic. There was no way, I thought, that we would let this happen again, that we would completely pull out of a country in disarray. *Not our problem*? Well, okay, maybe so. But we had made it our problem for twenty years, and it wasn't going to go away overnight.

What a shock it was, then, to witness such a total and complete disaster as American forces completely withdrew from Afghanistan in the summer of 2021, resulting in a swift and complete collapse that left the Taliban in control of the country and cost the lives of thirteen American soldiers and scores of Afghan civilians.

It was kind of like watching a train wreck in slow motion, with each day worse than the one before it. Everyone on the sidelines had been mistaken. We all figured that somebody—from the president on down to the generals who were advising him—would pump the brakes and realize that it wasn't just a military crisis happening in Afghanistan, it was a humanitarian disaster. Even into August, as the mobs of desperate civilians descended on Hamid Karzai International Airport in Kabul hoping to escape retaliation from the newly installed Taliban rule, I kept waiting for the right decision to be made, for an infusion of American troops and special forces that would at least slow the descent into madness.

It never happened. Instead, things got much worse before they got better. Actually, things never got better, at least not for those left behind. For those of us who served in Afghanistan, it was an enormously sad and frustrating time, but we did what we could to help, even as life went on back home. Amid one of the busiest fire seasons on record, we were drawn into the chaos as the withdrawal of U.S. forces—and the bedlam that ensued—stranded tens of thousands of Afghans whose safety was jeopardized because of their resistance to the Taliban and support of the U.S. military during two decades of conflict. The withdrawal, which quickly became a chaotic evacuation, caused constant drama for those of us at Bridger who worked long hours flying into and around fires all day and then turned to working the phones at night, helping to assist desperate families who were trying to get out of Afghanistan, many of whom we had known for years and with whom we had close working relationships and friendships.

I'm not sure many Americans realize the bond that U.S. soldiers formed with some of the Afghan soldiers and civilians who assisted our mission, often at great risk to their own lives. Regardless of whether you feel it was or wasn't our war to fight, it had become our war, and we owed a debt not only to the American military professionals who remained there in the harrowing closing days, but to the many Afghans who supported us. The idea that we were suddenly abandoning them to whatever fate the Taliban deemed appropriate (and we all knew what that would look like) was morally repugnant.

I had spent nearly eight years helping my former interpreter achieve safe passage to the West. Fida was a well-educated man who had been a pharmacist before the war. And by "war," I am referring to the massive escalation of 2009, when our military footprint in Afghanistan went from roughly four thousand troops to nearly one hundred thousand practically overnight. It was then that our involvement evolved from an exercise in keeping terrorism at bay to the herculean task of transforming Afghanistan into some semblance of a democracy, just as we had attempted to do in Iraq. We were toggling between two wars, and for the time being, at least, our emphasis was on Afghanistan. It was a totally ridiculous strategy that had no chance of success, but that was the mission, and those of us charged with fighting did the best we could under difficult conditions against an enemy that had literally spent centuries turning back every occupying force that had crossed its border.

More than anything else during this surge, we needed the assistance of qualified and reliable linguists, because, after all, not many Americans speak Pashto or Farsi, or any of the half dozen languages and dialects that we would encounter daily. As anyone who spent any time in Afghanistan quickly realized, once outside the wire, your most valuable asset beyond your weapon was a skilled interpreter—not just someone who spoke the regional language and a bit of English, but an intelligent and trustworthy linguist who could toggle fluently between the two languages and communicate the nuances of frequently tense conversation. The only people qualified to do this work were highly educated professionals—doctors, lawyers, teachers, pharmacists—who either had lost their jobs during the war or who were willing to risk their lives for compensation that was, in many cases, more money than they had earned in their previous careers. The most skilled linguists, those who worked for American intelligence agencies such as the CIA, were often paid well into six figures. Many of these interpreters had already left Afghanistan and held jobs in the U.S., Europe, or Canada. But when presented with an opportunity to earn significantly more money working as interpreters in Afghanistan, they returned to their homeland for months at a time despite knowing that by aiding and abetting the enemy (the United States), they had been marked for death by the Taliban.

Slightly lower in the linguist pecking order were the equally well-educated professionals who, for the most part, still lived in Afghanistan and had not worked or traveled extensively. These interpreters were also paid quite well and often assigned to Special Forces. Fida fell into this category, and the only difference between him and interpreters at the highest level was that he lacked the cultural pedigree that comes with having lived in London, New York, or San Francisco or having a degree from Stanford or MIT or another elite American university (which some of the linguists possessed). That aside, he was extremely capable. He was also, when circumstances dictated, a fighter, which is more about heart than skill, but when clearly demonstrated earned the linguist profound respect from his American colleagues.

For an unemployed pharmacist-turned-linguist, Fida was a warrior. On one occasion, a member of our team was wounded in a firefight, and Fida ran out into the open and pulled the guy to safety, at great risk to his own safety. He was never viewed as anything less than part of the team, which was not true of every interpreter. We had more than a few linguists that were generally despised, despite their skill, because they clearly were in it only for the money. Now, obviously we all understood the difference between soldiers and non-soldiers, between linguists and Special Forces operators—no one expected an interpreter with no formal training to suddenly perform like a member of Delta Force or the SEALs or any other member of the American military. Frankly, we did not even expect them to perform like members of the Afghan military, some of whom were quite courageous and skilled, and some of whom were not. But we did expect and appreciate acknowledgement that the job came with certain risks, and that there would be times when we would all be thrown into the shit together. Some of the interpreters were lazy and frightened. They wore sneakers despite knowing they would be hiking in the mountains for days on end. When the shooting inevitably started, they'd run and find the nearest and deepest hole and not emerge until the last bullet had been fired. They weren't really part of the team. They were mercenaries—in the worst sense of the word. But there were other guys who believed in the mission and became part of the team. They cared and they fought.

Fida was one of those guys.

Eventually, though, he saw the writing on the wall and knew he had to start planning for a future beyond Afghanistan. To most of the Afghan interpreters, a job in the United States or Canada was the golden ticket—access to a new and better life. But as Western forces steadily withdrew from Afghanistan and Taliban influence regenerated, there was a sense of urgency for those who had assisted the U.S. mission. It wasn't merely about finding a better life in America; it was about staying alive.

My last deployment ended in 2014, and I detached from the Navy later that year, but I remained in touch with Fida even as I transitioned to civilian life. Like so many others in his position, Fida spent much of his time jumping through bureaucratic hoops in the hope of getting out of Afghanistan and coming to the United States. It wasn't enough that he had served both his country and ours. He had gathered a mountain of evidence demonstrating his worthiness: letters of recommendation and sworn statements from officers and other supervisors who had worked with him during various assignments over the years as well as affidavits supporting his character and skill, essentially saying, "Hey, this guy is not a terrorist or a thief—he's a good man and he should be allowed to come to America (or Canada, or wherever he wants to go)."

I wrote countless letters on Fida's behalf and helped him put everything together. Eventually, in 2015, he got to the point where he was ready to apply for a visa. But as so often happens with bureaucratic processes, things did not go smoothly. There was, you might say, a glitch in the matrix. Most interpreters worked for periods of several months at a time, after which they would return home for a while, and then, if they were so inclined and their performance was laudable, return for repeat assignments. It was not unlike the process of deploying in the military. Unfortunately, one of the requirements for obtaining a visa involved sustained work over a period of time. A gap of more than thirty days required a "reset" of one's service clock. This happened to Fida when he returned home for a while to help his family out. Additionally, on one of his assignments, Fida was attached to a Canadian military outfit for roughly nine months, and of course the Canadians and Americans document things differently. Even though the work was basically the same, and even though the U.S. and Canada are the oldest of allies and fought alongside one another in Afghanistan, Fida was informed that his Canadian service would not be considered in his application for a U.S. visa. It was ludicrous. Fida had no say in where he worked or with whom he worked. He was hired by a private contracting firm that assigned him to military units based purely on circumstance and timing—he went where he was needed, when he was needed. He certainly never expressed a preference for working with the Canadian military rather than the American military. He did what he was told to do, and he did it well.

Regardless, these various issues prevented Fida from obtaining a visa for several years, despite having significant support from friends and former colleagues in the U.S. Finally, in 2020, we were able to help Fida and two of his older children get out of Afghanistan. They first went to Northern California before eventually settling in Chicago. Unfortunately, Fida couldn't bring his entire family with him at the time since there were different rules governing visas for his wife and kids, and two of the children were quite young, so they stayed behind with their mother. The plan was for Fida to bring the rest of his family over a few months later, after he was settled. But a process that was already dysfunctional became utterly chaotic as 2020 gave way to 2021, and the U.S. withdrawal from Afghanistan dramatically accelerated. Suddenly,

Fida was faced with the very real prospect of never seeing his wife and younger children again.

This is but one story (and I'll come back to it in a moment) among thousands, but it perfectly illustrates the human consequences that accompanied the end of our involvement in Afghanistan. As Bridger Aerospace is a company with deep military roots and many employees who put boots on the ground in Afghanistan, we watched with great distress as the drama unfolded horrifically, tragically in the summer months. But we also took action, as did many others.

August 2021 saw the Taliban regain control of Afghanistan as Western troops withdrew, and with that exodus came some of the greatest suffering of the Afghanistan conflict, as tens of thousands of civilians and former military personnel sought to leave the country rather than live—or die—under Taliban rule. The U.S. was bound by its commitment to withdraw its troops by August 31, and Western countries were left largely unprepared to execute the safe and swift evacuation of those in need. Operation Sacred Promise was born out of this humanitarian crisis. Comprised of over two hundred volunteers, the organization offers wide-scale evacuation and resettlement assistance to our allies from the Afghan Air Force and Army Special Mission Wing and their families.

Among those who benefited from the extraordinary efforts of Operation Sacred Promise was a thirty-two-year-old Afghan pilot named Samimullah Samim. I did not know Sam at the time, but his story of struggle and survival—as well as his commitment to working alongside American troops in Afghanistan—came to my attention through retired Air Force Lieutenant Colonel Ryan Cleveland, a pilot at Bridger. Ryan had joined Bridger in 2020, following a stellar twenty-one-year career in the Air Force, lastly as commander of the 81st Fighter Squadron out of Moody Air Force Base in Georgia. As part of his duties, Ryan had overseen the training of more than one hundred Afghan pilots and mechanics who then returned to Afghanistan, where they served in the Afghan Air Force. While the training mostly took place at Moody Air

Force Base, the pilots also spent several weeks in Colorado during the summer, where conditions allowed them to get some of the high-density altitude training that they would need to be successful while flying over the mountainous terrain of Afghanistan.

As a result of that work, Ryan developed friendships with several Afghan pilots, some of whom began reaching out to him in the summer of 2021 as the Taliban takeover grew imminent.

"By the middle of August," Ryan recalled, "I was getting texts regularly from some of the people I knew over there. They were all saying, 'What do we do? How do we get out of here?'"

That simple question had no simple answer, but it resulted in the formation of Operation Sacred Promise, a grassroots nonprofit organization that developed quickly under the guidance of Chief Executive Officer David Hicks, a retired Air Force brigadier general; President Carl Miller, an Air Force command pilot who was squadron commander for all flying operations in Kabul; and Chief Operating Officer Chris Ibsen, who served most of his career with Air Force Special Operations Command. The organization was, and continues to be, guided by three objectives:

1) The safe and lawful **evacuation** of at-risk allies and their families from Afghanistan.
2) The **resettlement** of these families in the U.S. or other countries by assisting with their legal status, housing, employment, and social integration.
3) Legal and policy **advocacy** to ensure that Afghans are granted full and timely access to the U.S. immigration system.

As the Taliban gained control of every major city in Afghanistan and anxious evacuees flooded Kabul Airport, the most pressing of these goals was evacuation. Just how pressing depended on an individual's particular situation and background.

Samimullah was a specialized pilot in the Afghan Air Force. He was an attack pilot, which meant he was directly involved in the fight and, as such, responsible for the elimination of Taliban forces. There were three groups in the Afghan structure that were marked for death by the Taliban. The first were the interpreters because they were considered the ultimate traitors—collaborating directly with the enemy in exchange for money. Interpreters were not merely killed; if captured, they could expect to be killed in a terrible, often public manner. They would be tortured, decapitated—sometimes in front of their families. And sometimes, their families were killed in front of them before they were executed. Heinous, barbaric stuff.

The next group down from the interpreters were the commandos—the special operations soldiers that were trained by Western forces. The regular Afghan Army guys who fought against the Taliban were mostly given a pass. They were working stiffs trying to feed their families. They were not necessarily "true believers" in the Western mission and therefore were granted a degree of compassion or understanding by the new Taliban regime. But if you were part of Special Forces, well, that was different, both in terms of perception and duty. The Afghan Special Forces soldiers had gone through some tough high-level training, mostly under the direction of U.S. Special Forces. These guys were, in fact, believers, and the Taliban knew it. They were openly and demonstrably pro-American, and the Taliban made it a point to hunt them down and execute them.

The third group marked for death were the pilots in the Afghan Air Force. The Taliban absolutely hated the pilots, who would come to the United States for two years or more and learn how to fly at an extremely accomplished level and then go back to Afghanistan and support the mission from above. They were not just traitors, but traitors with lethal capability—the devil in the sky.

All pilots were despised by the Taliban, but attack pilots, like Samimullah, were deemed the worst of the worst. He wasn't flying helicopters or transporting cargo; he was flying an A-29 attack plane—strafing the Taliban with laser-guided rockets and a .50 caliber machine gun in each wing, all in support of friendlies on the ground. He was marked for death and he knew it, and therefore, it was imperative that he and his family get out of the country as quickly as possible.

"Sam was one of the best students, if not the best student, that went through the program," Ryan recalled. "I have fond memories of him as a student, but I also looked up to him as a person. We had kept in touch a little bit, so I knew the struggles that he was going through back in Afghanistan. Obviously, we wanted to help him as much as we could."

Sam had an unusually strong pedigree. The son of a colonel in the Afghan military, he had first come to the United States in 2010 for a military competition at West Point. He was a student at a military academy in Afghanistan at the time.

"I had never even seen the inside of an airplane before that trip," Sam remembered with a laugh. "I had no idea I was ever going to be in the air force."

Upon completing his academy training, Sam was selected for an aviation leadership program, which brought him back to the U.S. for training and language education. He graduated in 2014 and entered the Afghan Air Force. Logistical and paperwork issues (not unusual in Afghanistan) forced Sam to leave the military and find work elsewhere to support his family for a while, but eventually, he was recruited back into the system and joined Ryan's A-29 program in Georgia.

"I got lucky," Sam said. "Some of my classmates told Carl Miller that there was a pilot who had gotten lost in the system and was trained by the U.S., and now he was out there doing nothing. So, they found me and reached out."

Sam was a talented and determined student who even survived an ejection from his plane during a training mission that left him with chronic back pain. Nevertheless, he served with distinction for many years in the Afghan Air Force, flying A-29s until late 2019, when he became the director of a program training a squadron of caravan attack pilots.

"I wanted to be a good pilot. I wanted to improve myself in the system, and I had a feeling I could do big things for the country," Sam explained. "I always wanted to get to places where I could do bigger things—where I could do more to help. There were things I loved about the military, and there were things that needed changing, and I felt like I was on track to helping with some of that. Obviously, I wanted to support my family. I wanted to have my own house and provide for my wife and children. I wanted them to have a good education and a better life. For the country—I just didn't want to lose what we had and what we had fought for the last twenty years. For me, personally, part of my job was helping the U.S. Embassy. I risked my life to give them information, so obviously, I didn't want the U.S. to leave because I didn't want our government to fall into the hands of the Taliban. I knew that would be terrible for my family, and for the whole country."

By the summer of 2021, Sam was working long hours with both the training program and as a pilot, flying medevac helicopters, rescuing wounded soldiers from Helmand Province. On August 14, he woke at his home in Kabul to discover that the city had fallen to the Taliban.

"My father was visiting," Sam recalled. "He said, 'The government has collapsed. We need to make a plan.' But it was a very bad situation. We had no passports, no preparation for anything like this. We were trapped. For about ten minutes I was just shocked; I thought it was a dream, like I was having a nightmare or something. It just didn't make sense. I woke up and went outside and saw, like, the light of the sun, and I just stared at it. You know when you see the morning light and sometimes it just looks…*different*? That's what it was like. It looked like the end of the world."

Figuring logically that he was putting his family in danger by remaining in the house, with the Taliban bearing down, Sam first went to stay with a civilian friend while he formulated a plan. That night, he heard from Carl and Ryan, who advised him that he should try to get to the airport and contact American soldiers who were serving as gatekeepers, charged with controlling the swelling mass of people attempting to escape.

"Initially, I thought it might be a good idea for me to try to get out alone because it was so dangerous in the city and around the airport. I thought my dad could take care of the kids while I tried to get out and then I would send for them, and they would be safer traveling without me. But that just didn't feel right. There was so much unknown and things happening so fast. I decided that I would try to go to the airport first, alone, to see what it was like. I knew that going through all the checkpoints would be dangerous, but at least this way if I got killed, I would be killed alone. And if I made it, then I would draw a map in my mind, detailing all the checkpoints and how to deal with them when I had my wife and kids with me. Honestly, I did not think it would work. I was prepared to die."

The checkpoints were mostly manned by young Taliban soldiers, some not more than sixteen years old, armed with automatic rifles and charged with stopping and interrogating everyone who was trying to get to the airport. If Samimullah's identity were revealed, he would have been executed. So, he did not bring an ID with him on his journey to the airport, instead relying on his powers of persuasion combined with the fact that the Taliban guards were overwhelmed by the sheer number of people at the checkpoints.

"I told them I was a teacher and that I had lost my ID," Sam said. "At some checkpoints, it was easy—they just let me go through. At others it was not so easy. One kid started slapping me around, saying, 'I know you're lying!' But he couldn't prove anything or find anything on me, so he just gave up. I thought he was going to shoot me. For them, it was easier to just kill people than to ask a lot of questions and go into details. They didn't have time, and there were too many people."

Sam went to the airport twice on his own to assess the viability of his plan before finally taking his wife and children with him. But this was far riskier, as he would need his ID to gain access to the airport and help from American soldiers. And yet, he could not reveal his identity to the Taliban guards along the way, for they surely had his name on a list of Afghans who were marked for death due to their support of Western forces. They decided to place all their belongings, including clothing for three children ages five and younger, along with Sam's identification records, in a single backpack that would be carried by Sam's wife. The hope was that at any given checkpoint, the backpack would be deemed benign, and Sam's family would be ushered safely through.

Remarkably, that was the way it worked, although the final hours of the journey were enormously intense, as Sam and his family encountered massive crowds at Kabul Airport. Sam stayed in contact with members of Operation Sacred Promise, who urged him to remain outside the airport for more than twelve hours, rather than returning to his home, and then guided the family to a particular gate and spoke by phone with a Marine who reviewed Sam's ID and allowed him inside the gate. Several hours later, he and his family were on a plane bound for Qatar, then another flight to Germany, and finally, a week later, a flight to the United States.

"Right until the very end, I felt pretty hopeless," Sam remembered. "I still sometimes can't believe we got out."

Not everyone was so fortunate. Less than two weeks later, on August 26, 2021, a suicide bombing at the airport killed thirteen American service members, including eleven U.S. Marines, a Fleet Marine Force Navy Corpsman, and a U.S. Army soldier. At least 170 Afghan civilians also were killed in the attack; another 150 were injured, including American military personnel, Afghan civilians, and Taliban forces. It was a devastating loss of life and a crushing denouement to the United States' twenty years of military engagement in Afghanistan. The average age of the American soldiers who died that day was twenty-two. I recognized them as brothers, as profoundly courageous servicemen who selflessly made the ultimate sacrifice. But now, approaching my late thirties, a veteran of the Afghanistan war myself but with four small children, I recognized them as something else as well: they were kids, barely beginning their adult lives.

For me, and for many American veterans, it was one of the worst days of my life. Regardless of how you felt about the seemingly endless conflict in Afghanistan, it was impossible not to feel both shame and sadness at the way it came to an end. I blame not only our political leaders, but the generals who advised them on both sides of the aisle. Those guys get paid to make hard decisions and provide good advice, and they completely screwed this up. I tried to capture some of what I was feeling in an email I sent to the Bridger family, many of whom were veterans, a few days after the bombing.

Teammates,

I haven't sent this note out sooner because, frankly, I didn't know what to say. To be honest, I still don't know exactly what to say, I have never felt the cocktail of emotions that I feel now about what has happened and is happening in Afghanistan. I am a loyal American Patriot, loyal to our allies in Canada, Australia, UK, NATO, and our Afghan allies. I am always a team player who rides for the brand. But seeing what my brand has done in the past several weeks has left me questioning many things, things I never thought I would have to question.

For those of you who didn't serve, you may have a curiosity about what is going on there and be trying your best to make sense of what you're seeing on the news, and most likely coming to conclusions that align with your political orientation. However, for you, the feelings are going to be less raw and contorting to your soul.

For those who did serve in Afghanistan, lost friends, lost family members, engaged in fierce and unforgiving combat, missed birthdays and anniversaries, delayed your careers and lives to serve a higher cause, I don't have any comforting words to share with you yet. I don't know what to say to myself or others about what we did, what we saw, and what is happening now. We were attacked, thousands of Americans were murdered, and we did the best that we could with the information we had at the time, we reacted, and many of us answered the call to serve. As I read the countless news articles by academics and pundits who never served a day in Afghanistan, I feel as though I am bouncing between rage and befuddlement at how misunderstood the conflict, the people, and the war have been. I have made many attempts at writing op-eds, letters, posts, or emails trying to explain to politicians and citizens all the wrong things they think they know about Operation Enduring Freedom and all the associated efforts. But every time I try to write those articles, I start to realize what most combat vets have dealt with for millennia—any attempt to explain what happened and most importantly WHY it happened will be a gargantuan task and ultimately fruitless for the speaker and the listener. You simply become comfortable with the view that "they will never understand."

The best solace I can give you is this: Millions of Afghanis are risking their lives, climbing mountains, enduring beatings, falling from aircraft trying to come to America or the West and build new lives here. We showed them, for 20 years, that the Western spirit and the freedoms that we all take for granted are rare and ought to be cherished. Our domestic definition of oppression pales in comparison to the daily life of a young girl in Kandahar province or a child in Nuristan. China nor Russia will present them the opportunity to live free and thrive as we did. As I have spent nights on the phone talking families, wives, and children through their escapes from Afghanistan, I am reminded that although we committed a terrible atrocity in pulling out how we did, an atrocity for which we will never be forgiven, we also spent 20 years showing the Afghan people what a free life looks like. They have now tasted that freedom and are willing to risk death and cross the world to find it again. If nothing else gives you solace, perhaps that will.

For any Afghan vets on campus or in the family who want to talk, I am always available. I won't be able to offer you words of encouragement, but I can at least lend an ear and hopefully ease your burden. For others who just have questions about what you're seeing and you want to understand more, you are welcome to reach out and we can have a conversation.

All my best,
Tim

In the ensuing days and weeks, many of us continued to work diligently to try to facilitate the safe passage of our Afghan friends and their families.

"It was absolutely heartbreaking," remembered Ryan, who, like several of us at Bridger, found himself flying long hours on fires during the day and then assisting with evacuations a world away in the evening hours. "It was the worst-case scenario playing out in front of our eyes—the idea that our friends would potentially be left there and then hunted down by the Taliban. We knew that the Taliban were going to be able to get into the old squadron buildings and figure out names and missions. The A-29 pilots were wreaking havoc on the Taliban, so we knew that they would be one of the number one targets. So, we obviously were extremely motivated to do whatever we could to help them. We had some advisers that were still in the military that told their commanders, 'I am doing nothing else for the next two weeks but this.' And they would literally be up all hours of the night and day, working with us to make sure everybody that we could get out actually got out."

In the wake of the tragedy, there were, improbably, a number of success stories. I spent weeks working with military buddies and contacts at the U.S. Embassy in Afghanistan to secure safe passage for Fida's wife and children, trying to guide them through checkpoints en route to the airport, much as Samimullah had done a few weeks earlier. Eventually, we got his family out by accessing a mobile route through Pakistan—and it happened despite getting very little help from the American government. It was the Canadian Embassy that stepped in and facilitated the necessary paperwork, which was ironic since Fida only worked for the Canadian military for a few months, as opposed to several years of work for the U.S. I found that disappointing, but at least they got out and were safely reunited with Fida, which, in the end, is all that matters.

Samimullah and his family, meanwhile, spent two months at Fort McCoy Air Base in Wisconsin, living among some fourteen thousand Afghan refugees who were waiting for placement in the U.S.

"That was not easy," Sam recalled. "There were six families staying in one small house. We used curtains to create space and some privacy, but it was challenging. The kids would cry a lot, and some people were in bad condition emotionally. But we knew we were lucky to be alive and together."

Through Operation Sacred Promise, Sam and his family made their way to Florida, where they lived with a host family. A fundraiser helped raise thousands of dollars for Sam to attend flight school and take written exams so that he could begin the long process of acquiring the necessary credentials and certifications to fly in the U.S. The family later moved to Cleveland to be near some Afghan friends while Sam received more instruction at a flight school in New Mexico. While this was going on, I got a call from Ryan, who told me all about Sam.

"This guy is here, he's a terrific pilot, and he's doing the training to get back into aviation. Any chance there's a job for him at Bridger?"

I didn't hesitate before responding.

"Absolutely. He sounds like he'd be a great hire. Let's find a way to make it happen."

Our VP of business operations, Sam Davis, one of our top leaders and performers in the organization, worked hard with our rock star general counsel, James Muchmore, to find a way for Sam to become part of our team. Samimullah joined Bridger as a pilot in May 2022, where his primary responsibilities were fire mapping and transporting equipment and supplies to various fire bases in the West. When he came to Bridger, he knew almost nothing about firefighting—in fact, he was not even aware that wildland fires were a significant problem in the United States. But he was a quick learner, and like so many of us who have joined the fight, he soon developed a passion for work that is best described as transformative.

"Sometimes, when we stop a fire," Sam explained, "it gives me the same feeling as when I was flying back home, supporting the friendlies when they were ambushed or attacked by the Taliban. When I would defend them and they would be super happy afterward, when I would talk to the commander, and I would have this feeling like…I'm doing something not only for myself, but for my country, for the people. Firefighting is the same kind of thing. It has meaning. I feel like this is what I'm meant to do."

EPILOGUE

The aerial firefighting business is booming. Make of that what you will.

Bridger Aerospace continues to grow and expand at a rate that I never dreamed possible. To think that the company we started in 2014 with just a handful of dreamy optimists who knew almost nothing about wildland firefighting is now a billion-dollar business with over three hundred employees is mind boggling. I am proud of the work that we have done and the service that we provide. I would like to believe that our success is a byproduct of hard work and commitment and progressive, innovative thinking in a business that for too long was stuck in the past. We've been fortunate to hire a lot of talented people—pilots, mechanics, engineers, sales, and marketing personnel—all committed not just to making Bridger a successful company, but to fighting the extraordinary threat posed by an ever-expanding wildlife season.

But therein lies the conundrum. As with any business that relies on the continuance of bad things happening, Bridger's success—or at least its opportunities—stems directly from the fact that wildfires are not just a persistent problem throughout the world, but one that has grown exponentially in the last decade and shows no sign of abating. For all the reasons we've already discussed—climate change, wildland-urban interface, decades of accumulated fuel, and countless other factors—wildfires represent one of the greatest threats to modern civilization. The management and mitigation of that threat is, in my opinion, among the more important pursuits one can undertake in the twenty-first century.

★ TIM SHEEHY ★

My intention in writing this book was principally to bring the story of aerial firefighting, its heroes, and its warhorses to a wider audience than currently exists. Inevitably, it also ended up being a story about my entry into the industry, and hopefully that provided some enjoyable color along the way. Moreover, I wanted to use this platform as a call to action—a vehicle for provoking discussion and interest in a problem that is too often ignored or underestimated, one that gains national attention only when disaster strikes on an epic level; and even then, it is quickly forgotten. Given the current state of wildland fire management, I am not worried about running out of work. I would just like the tools and the support to do the job as well as it can be done. Most importantly, I want the heroic men and women in the aircraft who risk their lives to be recognized and appreciated by the people whom they protect. Nothing can help do that better than telling their story.

The U.S. posture for wildfire prevention, management, and suppression remains grossly under-resourced at all levels. Our collective appreciation for the impact of this threat is generationally out of sync with reality. We must have a nationwide force that can respond quickly, aggressively, and appropriately to any and all wildfire incidents within thirty minutes. If you tell a citizen that their home could go seventy-two hours without a fire engine or police officer responding to a 911 call, they would be apoplectic. But that is the nature of wildfires today in the U.S. We can go *days* without an aerial response, and this is unacceptable. All agencies with a responsibility for wildfire suppression and mitigation are routinely underfunded given the scope of the problem they face. Consider that approximately four billion dollars was devoted to wildfire-related federal spending in 2021. That might sound like a big number—until you realize that the economic impact of wildfires in that same year was estimated to have exceeded four *hundred* billion dollars! And that does not include the long-term health effects of smoke inhalation and related illnesses (to both firefighters and citizens who live in fire zones), carbon dioxide emissions, or longer-term economic implications such as rising homeowners' insurance, displacement of workers, and emotional distress.

Compare that four billion dollar budget to an annual defense budget that typically exceeds $750 billion, and it's hard not to think that something is amiss. As a nation, we can do better. I say that as both a proud veteran of the U.S. military and a first responder in the battle against wildfires.

As a society, we are constantly evaluating incremental or plateauing steps to arrest the spread of a problem that—pardon the overdue pun—is spreading like wildfire. Denton, Montana, in December; Denver in January; Florida in February. Nova Scotia and Quebec in June, sending walls of smoke to Boston and Baltimore—even as far south as Atlanta! A decade ago, none of this would have been considered "normal" wildfire activity. And yet, it is becoming increasingly commonplace. Year after year, the impact of wildfires around the world—from California to Sweden—pushes the boundaries of global sustainability. There are not enough firefighters, not enough aircraft, not enough equipment, and not enough off-season education and preparation. Some say it's a funding issue. I disagree. There is plenty of funding available to the wildfire problem—it's just allocated to so many of the wrong budgets. We have vast sums of money reserved for "suppression funds" (think emergency funds) that when spent, cost us three times as much as it would if it were allotted to the preparedness budgets. We underfund the people and equipment we need to be ready…and overfund the kneejerk reaction to put out the fires. While it's easy to cast blame on the agencies and governments that control the purse strings, that is a short-sighted and ill-advised response. It is our fault, as citizens and taxpayers and industry stakeholders, for not holding our elected officials accountable for underfunding and under-resourcing our wildland management apparatus.

The impact goes well beyond the burning of houses, although that certainly tends to garner the most attention. There is a direct correlation between carbon emissions produced by wildfires and the ever-intensifying choking of our atmosphere by greenhouse gases. Wildfires raise insurance rates or prevent homeowners from getting insurance at all. This, in turn, affects mortgages and the insurability of other assets at the citizen's home, thereby increasing the cost of living and the risk of living in a fire-prone area. It further impacts economic recovery in communities at risk or previously affected by wildfire.

As I evaluate this paradigm through the lens of someone relatively new to the industry, I am baffled and concerned. In the military, we would spend millions—or even billions!—of dollars preparing for even the slightest eventuality of a threat materializing into an actual event. We would prepare, train, drill, train some more. Money was no object when balanced with the threat we faced.

Yet as we examine the harsh and inescapable realities of the societal threats posed by wildfire and its causes, the funding pool remains static or marginally increased. From forest thinning to ground firefighter staffing and pay, from fire equipment to investment in fire bases, we simply do not allocate enough money to get the job done. Coupled with an inflationary environment and a global aviation labor shortage, the next decade does not bode well on our current trajectory.

The time to press for change is now. A dollar spent on prevention saves a thousand dollars spent on recovery. We don't have to lose communities every year—government agencies have the leadership and strategy in place to better prepare for a full-spectrum response. But without significant increases in funding for suppression assets on contract, we will be in a losing position. It will take years, if not decades, to overcome the challenges we now face, and frankly, I don't see much progress moving the center of gravity in the right direction.

Technological innovation can mitigate some of these issues, and it remains a powerful force within the industry. Bridger Aerospace has entered more aircraft into fire service than any other provider in the past several years. Our obsession with innovation in support of saving lives drove these investments and continues to fuel our advances in all areas related to aerial firefighting. We continue to explore new airframes, technologies, and tactics that can enhance our ability to save lives and protect communities.

To that end, in 2022, we launched a groundbreaking free mobile application called FireTrac. The culmination of two years of research and development, at a cost of several million dollars, FireTrac represents a melding of our experience developing sensor systems for the military and providing aerial services for our firefighting community. FireTrac provides real-time push notifications to any subscriber, along with rapidly updated fire maps collected and shared by our fleet. Its purpose is to democratize fire information that has previously only been available to select government agencies, making that information available to any citizen who desires it. With a friendly user interface, FireTrac scrapes information available from public sources to compile an up-to-the-minute mosaic profile of each fire incident. With fire line mapping and active fire imaging overlaid for user consumption, this creates a unique and potentially lifesaving tool for citizens in fire-prone areas. Moreover, FireTrac, which is active year-round, will enable agencies, landowners, property managers, and concerned citizens to rapidly quantify the impact each fire has had on a local ecosystem and adjust management practices accordingly.

Information and intelligence sharing in the American wildfire realm is woefully inadequate. Citizens threatened by fires usually piece together an information mosaic from a myriad of different sources: Facebook, Twitter, local news, text strings from friends, maybe a government website. Most of these sources are not data aggregation specialists or wildfire experts. They are just regular citizens trying to make it through a devastating and terrifying event. As the Paradise fire in 2018 taught us, entire communities can be caught off guard by a fast-moving fire. We can do a better job of putting information into the hands of those who need it most. FireTrac is an attempt to answer that call. It won't be perfect, but if it can save one life, or one house, then it will be worth it.

The next generation of fire innovation will be data and intelligence. We must get better at gathering, analyzing, and disseminating critical wildfire information. This includes wildfire imagery, real-time location data of firefighters and equipment, aircraft location, firebreak and fireline status, predictive analysis with weather overlay, and, of course, the ability to seamlessly update and share this data with as wide a set of stakeholders as possible, including local citizens. Sharing wildfire information with citizens has long been a controversial topic with fire managers and government officials. They would rather no information get out than either bad information or information they don't want shared—for whatever reason. So, instead of working to actively share information with the population, they generally resist sharing until they deem it appropriate, which leads to a vacuum filled by DIY solutions. Although some local governments have made valiant efforts to fix this, since the federal government has such a strong hand in almost all wildfire incident responses, any data sharing effort without comprehensive federal involvement is largely pointless. Our hope is that FireTrac will inform and educate more people about the wildfire threat in their area, as well as the importance of what our community is doing to protect them—and above all else, encourage them to be actively involved in the process. Wildfire expansion, after all, is a crisis that affects all of us in one form or another.

Unfortunately, getting that point across has been, to put it mildly, a challenge, which is why aerial firefighting as an industry and the government agencies to which it answers must form a coordinated and passionate lobby. We must speak with one voice to those who craft our budgets, implement the strategies we deploy, and are responsible for our lives and livelihood. Indeed, aerial firefighting is different than almost any other industry in that it is us—the vendors and pilots—who put our lives on the line to accomplish this mission. We take the financial risk of acquiring, modifying, and fielding extremely expensive aircraft, and we take the mortal risks associated with flying those aircraft in hazardous environments to protect others. With that in mind, we should and must have a more meaningful seat at the table as the direction of our mission is determined.

★ MUDSLINGERS ★

Refer back to the first sentence of this chapter: *business is booming*. That is not cause for celebration but rather a clear-eyed statement of fact. And it's not going to change anytime soon. There is more than enough work to go around for all of us. The nation deserves one team fighting together to support our men and women on the ground and protect those most affected by the threat of wildfire. What does this mean? Well, for starters, the setting aside of petty grudges and territoriality so that we can employ all possible resources for a given fire incident. There are plenty of fires that merit the combined effort of a DC-10 and BAe 146, a fire boss and a scooper. One is not universally "better" than another; it depends on the type and location of a fire. Where a multifaceted approach makes sense, throw everything at the problem. If we coordinate our efforts as an industry, we can quickly provoke the kind of change we so desperately need to meet this mutating threat and evolve our capabilities deep into the twenty-first century.

Not long ago, I had the good fortune to address the North American Aerial Firefighting Conference in California. Part of my message was this: We are all in this together. Accountants, bookkeepers, mechanics, pilots, contract officers, salespeople, engineers—we are all fighting the same beast. I have worked hard to communicate this message to our staff—that when we are in the field, we must stand shoulder to shoulder with other providers to ensure we are collectively providing the best service we can to the firefighters and citizens on the ground. Yes, we are a business entity, but we are in the business of saving lives. Everything else is secondary. And we must embrace that mission and the mantle of responsibility that comes with it.

It is my hope that we can reconcile past competitive differences that pit LATs (large airtankers) against VLATs (very large airtankers), fire boss against scooper, retardant against water, and rotor against fixed-wing, and instead form an integrated advocacy group that speaks with one voice to our elected officials and the American people—who, after all, pay the bills. We must convince the citizenry that, as a team, we are the best hope for fire-prone communities to have a fighting chance when their number is up and they are faced with a fierce fire line approaching their doorstep. Wildfires present a clear and present danger to our national security, and it will take a full-spectrum approach to appropriately address it. A VLAT and a scooper working together on the same fire can be an immensely effective task force. Nighttime helicopters complementing a night ground assault will save lives. A year-round nationwide aerial response fleet has clearly become necessary. Forest management and prescribed burns are a critical part of fire preparedness and prevention, and they are initiatives that must be closely coordinated with year-round aerial supervision and suppression so that our national firefighting infrastructure is prepared, equipped, and postured to serve the people in need. Simply stated, we must be able to get to our fires faster, harder, and stronger.

As we learned during two decades of fighting in Afghanistan and Iraq, the most effective task force is one that embraces coordination across all functional areas to craft the most impactful effects on target. I am a relative neophyte to aerial firefighting and recognize that the industry has been around longer than I have been alive. But I can see that we are at a turning point for our community and the people we serve; if we want to have an impact on the future of aerial firefighting, the time to act is now. This is not to minimize the importance of reducing carbon emissions, being smarter about the types of building and materials that are permitted in fire zones, or carefully considering a more aggressive approach to forest management. All of these are crucial considerations in the battle against wildfires. But as a pilot and a businessman—and as something of a budding historian—my hope is that we can come together to advocate for improved solutions and resources to combat our common enemy: the loss of life and limb and property to the scourge of wildfires.

Walking the mountainous forests of the American West, one can't help but be in awe. We are a fortunate nation, in so many ways, but none more so than the pristine vastness of our forests, mountains, and plains. They are a national treasure and the spiritual embodiment of the manifest destiny that took America from an obscure experiment in constitutional self-rule to the most powerful nation ever seen. Billions of people depend on the cattle, wheat, timber, poultry, and metals that emanate from this region. For the people who have chosen the hard life of frontier living and the people who have chosen the luxury life of McMansions in the mountains, wildfire is a very real threat to their lives, their livelihoods, and their homes. For those who don't live in the fire zone, they will find that more and more, their fates are intertwined with those who do.

The twenty-first century will pose many challenges that call into question the very essence of modern humanity and what it means to be a responsible citizen of the world. For the first time in a generation, the words "world war" are again on our lips; economic upheaval and global pandemic have reordered many things. Wildfire is a small piece of this mosaic, but it's not too small to be ignored. Aerial firefighting plays a critical role in this arc and will require more support to protect those in need, which is rapidly becoming all of us. The time is now to usher in the next generation of the Mudslinger, because the biggest battles are yet to come.

GLOSSARY

I*n compiling the following list of commonly used terms, the author consulted a variety of sources that have published glossaries related to aerial firefighting and wildland firefighting, as well as accessing his own knowledge and expertise. The author would like to acknowledge, in particular, the U.S. Forest Service and Department of Agriculture, the National Park Service, the U.S. Department of the Interior Forests and Rangelands, the Associated Aerial Firefighters, the Texas A&M Forest Service, and the National Wildfire Coordinating Group.*

Aerial Firefighting – The implementation of aircraft in the support of wildfire management or suppression.

Air Attack – The supervisor in the air—typically in a fixed-wing aircraft—who oversees the process of attacking wildfires and directs all air traffic within the fire zone. Can also refer to the aircraft in which the supervisor travels, and that is piloted by someone other than the supervisor.

Air Tactical Group Supervisor (ATGS) – Senior official who coordinates incident airspace and manages air traffic within the fire zone. Sometimes used interchangeably with "air attack."

Air Tanker – A fixed-wing aircraft carrying water or retardant used for fire management or suppression.

Anchor Point – A location from which to start building a fire line that will serve as a barrier to the spread of fire.

ASM – Aerial Supervision Module. A lead plane with an ATGS qualified individual on board.

Backfire – A fire intentionally set along the inner edge of a fire line; goal is to reduce fuel load in the path of a wildfire or change the fire's direction.

Base (of a fire) – The part of the fire perimeter opposite the head.

Blow-Up (or Flare-Up) – A sudden increase in a fire's velocity or intensity caused by changes in weather patterns or fuel availability.

Break Left/Right – A command to turn left or right. Applies to aircraft in flight, usually while in the process of making a drop.

Brush Fire – A fire that is fueled predominantly by low-growth vegetation—shrubs and brush and small trees.

Bucket – A large hanging basket slung below a helicopter and used for dropping water on a fire.

Buffer – An area of reduced vegetation or other fuel that stands between a wildland (fire-prone) area and residential or commercial areas.

CAL FIRE – The largest state agency in the U.S. devoted to forestry and firefighting. Abbreviation for the California Department of Forestry, Fire Division.

Canopy – The uppermost layer of vegetation in a forest.

Chief Pilot – A pilot who has achieved a level of experience and training that allows them to act in a supervisory role within an organization.

Complex Fire – Two or more wildfire incidents within a region that combine to create one larger fire.

Containment – The process of creating a barrier or fire break around the entire perimeter of a fire, thus preventing further spread.

Control Line – A term for all constructed or natural fire barriers and treated fire edge used to control a fire's spread.

Conversion – The process of adapting former military aircraft for use in aerial firefighting.

Coverage Level – The density of retardant in a drop.

Crown Fire – A fire that advances primarily across the top of vegetation.

★ MUDSLINGERS ★

Direct Attack – The attempted suppression or smothering of a fire through the direct application of water or retardant, or by using ground crews to separate burning fuel from unburned fuel.

Dispatch – To send ground crews or aerial support into a fire zone. A dispatch is an official term that means ordering an asset to an incident.

Divert – To change an aircraft's assignment from one target to another.

Dozer Line – A fire line constructed by a bulldozer, as opposed to one that is constructed by ground crews using hand tools.

Drift Smoke – Smoke that has drifted from its point of origin.

Drop – An operation in which water, retardant, or cargo is released over a fire zone.

Drop Zone – The area around and immediately above the target to be dropped on.

Fingers – The long narrow portions of a fire projecting from the main body.

Fire Base – A base of operations established in the vicinity of a wildfire, staffed by personnel engaged in managing the incident; can also refer to a base for firefighting aircraft.

Firebombing – The dropping of water or retardant from an aircraft to assist with the suppression or management of a wildfire.

Fire Break – A strip of land from which the vegetation is removed for fire control purposes; any natural or constructed barrier meant to stop or slow the spread of a fire.

Fire Season – The hottest, driest time of the year, during which a geographic region is most likely to experience wildfire activity. In many parts of the world, fire season is lengthening and bleeding from one area into another.

Flanks – The parts of a fire's perimeter that are roughly parallel to the main direction of spread. The left flank is the left side as viewed from the origin of the fire, looking toward the head.

Fire Boss – A single-engine scooping aircraft, the AT-802.

Fire Head (or Head) – The most rapidly spreading portion of a fire's perimeter.

Fire Shelter – A small shelter of last resort, constructed of heat-resistant aluminum, that can be deployed when a firefighter is trapped within a fire zone.

Fire Storm – A wildfire of unusual intensity that creates its own weather system, marked by extreme updrafts and tornadic activity.

Foam – A type of fire retardant.

Fuel Load – The amount of burnable material present within a given geographic region; typically refers to dry vegetation (trees, shrubs, brush, grass).

Ground Fuel – Any material that will burn below the surface fuels. This includes leaves, moss, roots, rotting branches.

Hand Line – A fire line created by ground crews using hand tools.

Hotshots – A crew of highly trained ground firefighters dispatched into the most challenging wildfire scenarios.

Hotspot – A particularly active or volatile part of a fire.

Incident – A wildfire event that involves tactical intervention by firefighting crews.

Incident Commander – The person who oversees all tactical decisions at the site of a wildfire.

Infrared Detection – The use of infrared technology and thermal cameras in an aircraft or by satellite to find and trace fire activity that might not be visible to the human eye.

Initial Attack – The actions undertaken by the first responders at the site of a wildfire. This can refer to either ground crews or aircraft attempting to either suppress or control a fire event.

Island – An unburned area within a fire perimeter.

Knock Down – To quickly reduce flame in a specified area, usually by dropping water or retardant.

Large Air Tanker – A converted large, fixed-wing aircraft that carries over 2,000 gallons of liquid.

Late – Indicates that a drop was late or overshot its target.

Lead Plane – A fixed-wing aircraft that makes reconnaissance runs through a fire zone to ascertain conditions; also guides tankers to a target and supervises drops.

Low Pass – A low-altitude pass over the targeted area by either air attack or a lead plane to get a close look at the target, or to show a tanker pilot a target that is difficult to describe.

On Target – Acknowledgment to tanker or scooper pilot that his drop was well placed.

Origin – The place on the ground where a fire started.

Perimeter – The active burning edge of a fire or its exterior burned limits.

Prescribed Burn – A fire deliberately set and managed by professional firefighters to reduce fuel load and thus prevent a larger, more dangerous fire event.

Pre-Treat – Applying a retardant line in advance of a fire.

Probes – The small inlets that are extended into the water from a scooping aircraft to scoop water as the plane skims across the water. They are usually about the size of your hand.

Pulaski Tool – A tool used by firefighters for both chopping and digging, it combines an axe blade and a trenching blade attached to a straight handle.

Red Flag Warning – A term used by meteorologists to describe a weather pattern conducive to wildfire activity.

Retardant – A chemical substance dropped from aircraft onto a fire or at a fire's perimeter. Retardant can be used to directly suppress a fire or, more commonly, to create barriers that allow ground crews to manage a fire's spread.

Running – The behavior of a fire or a portion of a fire spreading rapidly with a well-defined head.

Salvo – Dropping the entire load of retardant at one time or dropping a combination of tanks simultaneously.

Scratch Line – A preliminary control line built with hand tools as an emergency measure to check the spread of a fire.

SEAT – Single Engine Air Tanker. A Thrush or Air Tractor aircraft converted from crop duster use with a fire-gate modification to drop liquid in one salvo on a fire instead of spray it on crops.

Secondary Line – A fire line built some distance away from the primary control line, used as a backup against spot fires.

Shoulder – Where the flank and the head of a fire meet.

Size-Up – An initial evaluation of a new fire.

Slop-Over – The extension of a fire across a control line.

Smokejumpers – Specially trained, elite wildland firefighters who help provide an initial response to a fire, usually by parachuting into or near a fire zone.

Smoldering – A fire burning without flame and spreading slowly.

Snag – Standing dead tree or part of a dead tree.

Split Load – The dropping of a partial load.

Spot Fire – Fire caused by sparks or embers that are blown outside of the perimeter of the main fire.

Spotting – When spots are being blown out of the fire line and causing spot fires.

Structure Protection – The act of doing targeting drops (usually from a helicopter) close to structures as a last resort to protect evacuating citizens and structures threatened by a rapidly advancing wildfire.

Surface Fire – A fire fueled mostly by smaller vegetation or loose debris on the forest floor.

Supe – The supervisor of a fire crew. The Top Dog for a ground team in the fight.

Super Scooper – A large fixed-wing aircraft that picks up water from a source near a fire and then drops its load directly on the flames. The only aircraft in existence built expressly for the purpose of fighting wildfires.

Suppressant – A substance, usually water or chemical retardant, that is applied directly to the actively burning portion of a wildfire.

Target – The object or area where water or retardant is meant to be dropped.

Traffic Pattern – The path of aircraft when landing or taking off.

Trail – To drop tanks in sequence causing a long unbroken line.

Type – A classification of resources and capability (human or mechanical). Type 1, for example, is considered the highest level of resource, followed by Type 2, Type 3, and Type 4. The ranking can be based on size, power, or experience and qualifications.

USFS – Acronym for the United States Forest Service, the largest federal forestry and firefighting agency.

Very Large Air Tanker (VLAT) – A DC-10. Carrying over 10,000 gallons of liquid. There was a 747 VLAT also in operation on and off for many years, but it did not gain wide industry traction.

Water Bomber – Any fixed-wing aircraft fitted with tanks that can be filled with water and deployed for use on a wildfire. Some water bombers are amphibious and can refill from nearby water sources without having to return to a fire base.

Wildland Fire – Any fire, other than a prescribed burn, that occurs in a natural setting. Distinct from structural fires, although wildland fires can also spread and affect structures.

Wildland-Urban Interface (WUI) – Refers to the growing phenomenon of civilization encroaching on areas previously deemed to be wild, creating zones in which wildfires can quickly spread to structures and communities with significant human populations. One of the biggest problems in modern wildland firefighting.

SELECTED BIBLIOGRAPHY & SOURCES

A*long with personal experience and numerous first-person interviews, the author utilized a wide variety of sources and published research in compiling the narrative for this book. They are duly recognized here.*

Aerial Staff. "A New Era for Air Attack." *Aerial Fire*, May 3, 2022.

Alexander, Linc W. *Fire Bomber Into Hell: A Story of Survival in a Deadly Occupation.* Booklocker.com, Inc., 2010.

Associated Press. "NTSB Points to an Existing Lack of Responsibility for Air Tanker Safety." December 7, 2002.

Atlas, Ted. "How an Ex-Hollywood Stunt Pilot and a Park Ranger Invented Aerial Firefighting." HistoryNet, May 4, 2022.

Atlas, Ted. "How the Death of Fifteen Firefighters in the Mendocino National Forest Sowed the Seeds of Today's Firefighting Aircrafts." MendoFever, June 13, 2022.

Barnes, James. "Remembering Joe 'Hoser' Satrapa." YubaNet.com, March 18, 2019.

Bridger Aerospace. Website: BridgerAerospace.com.

Bushey, Charles L. "Wildland Fire / Aircraft Firefighter Fatalities in the United States compared with Ground Based Firefighter Fatalities." Airtanker.org, December 2012.

Cermak, Robert W. and U.S. Forest Service. *Fire in the Forest: A History of Forest Fire Control on the National Forests in California, 1898 – 1956.* Government Printing Office, 2005.

Demerly, Tom. "Legendary Air Combat Pioneer, Vietnam F-8E Fighter Jock and F-14 Pilot Joe 'Hoser' Satrapa Has Died." The Aviationist, March 26, 2019.

Donati, Jessica. "Former Afghan Fighter Pilot Now Firefighting as Newcomers Join U.S. Workforce." *Wall Street Journal*, September 11, 2022.

Dyer, Clayton and Graham Flanagan. "How the $30 Million 'Super Scooper' Plane was Built to Fight Wildfires." *Business Insider*, March 18, 2021.

Egan, Timothy. *The Big Burn: Teddy Roosevelt & The Fire That Saved America*. Houghton Mifflin, 2009.

Ferguson, Gary. *Land on Fire: The New Reality of Wildfire in the West*. Timber Press, 2017.

Ford, Richard. *Wildlife*. Atlantic Monthly Press, 1990.

Frontline Wildfire Defense. "Aerial Firefighters & Fire Fighting: Dangerous… But Effective?" FrontlineWildfire.com, 2022.

Gabbert, Bill. "Environmental Group Files Lawsuit Against U.S. Forest Service Over Use of Fire Retardant." Wildfire Today, October 12, 2022.

Gabbert, Bill. "Joe 'Hoser' Satrapa – RIP." Fire Aviation, March 24, 2019.

Gee, Alastair and Dani Anguiano. *Fire in Paradise: An American Tragedy*. W.W. Norton, 2020.

Guillemette, Roger. "Aerial Firefighting." *U.S. Centennial of Flight Commission*, 2003.

Hildebrand, Richard L. "Firefighting Air Tankers, The Early Years." Fire Aviation, January 22, 2021.

Holpuch, Amanda. "U.S. Forest Service Planned Burn Caused Largest New Mexico Wildfire." *New York Times*, May 28, 2022.

Janney, Tom. "Airtankers: An Historic Overview." Airtanker.org, November 2012.

Jendsch, Wolfgang. *Aerial Firefighting*. Schiffer Publishing, 2008.

Johnsen, Frederick A. "Attacking Fires from the Air." *Air & Space Forces Magazine*, October 1, 2010.

Johnson, Lizzie. *Paradise: One Town's Struggle to Survive an American Wildfire*. Crown, 2021.

Kearse, Hannah. "Belgrade Company Brings Super Scooper to Montana." NBC Montana, May 18, 2021.

King Township Public Library. "RCAF Norseman Champions." Digital Archives.

Langan, Fred. "Canadian-Made Plane Used to Fight Forest Fires Scouts for New Uses." *Christian Science Monitor*, May 14, 1985.

Maclean, John N. *Fire on the Mountain: The True Story of the South Canyon Fire*. Custom House, 1999.

Maclean, John N. *Fire and Ashes: On the Front Lines of American Wildfire*. Henry Holt, 2003.

Maclean, John N. "Embers Forge Enduring Lessons." *New York Times*, October 31, 2003.

Maclean, Norman. *Young Men and Fire*. University of Chicago Press, 1992.

Malnic, Eric. "Two Air Tanker Crashes are Blamed on Fatigue Cracks." *Los Angeles Times*, September 25, 2002.

Martinez, Luis. "Inside the Daring Helicopter Flights that Rescued 200 from a California Fire." ABC News, September 7, 2020.

McDonough, Brendan. *Granite Mountain*. Hachette, 2017.

McLean, Herbert E. "The Great Airtanker Debacle." *American Forests*, May 1, 1994.

Meeker, Susan. "Rattlesnake Fire Marks 60 Years." *Tri-County Newspapers*, July 5, 2013.

Mooallem, John. "We Have Fire Everywhere." *New York Times Magazine*, July 31, 2019.

National Institute of Standards and Technology. "New Timeline of Deadliest California Wildfire Could Guide Lifesaving Research and Action." February 8, 2021.

National Transportation Safety Board. Various incident reports and statements. Ntsb.gov.

National Wildfire Coordinating Group. "First Airtanker Drop (California) – August 12, 1955." Nwcg.gov, August 2021.

NCW Life. "2021 National Wildfire Season was Longest and Most Destructive." Ncwlife.com, November 26, 2021.

Operation Sacred Promise. Website: opsacredpromise.org.

Popovich, Nadja and Brad Plumer. "As Wildfires Grow, Millions of Homes are Being Built in Harm's Way." *New York Times*, September 9, 2022.

Ramos, Jason A. and Julian Smith. *Smokejumper: A Memoir by One of America's Most Select Airborne Firefighters*. Mariner, 2016.

Richardson, S.D. "Operation Firestop." *Empire Forestry Review*, March 1959.

Santos, Fernanda. *Fire Line: The Story of the Granite Mountain Hotshots and One of the Deadliest Days in American Firefighting*. Flatiron, 2016.

Sierra Nevada Conservancy. "2021: Another Historic Sierra Nevada Fire Season." January 24, 2022.

Steffanson, Mike. "Nevada's Investment in Super Scoopers Proves Worthwhile in Avoiding Dramatic Wildfire Season." Kolotv.com, October 22, 2020.

Steller, Tim. "Aircraft in Scam May Have Flown for CIA." *Arizona Daily Star*, April 1, 1998.

Stellar, Tim. "Two Receive Prison Sentences in Military Aircraft Conspiracy." *Arizona Daily Star*, April 2, 1998.

The Backseat Pilot. "The Critical Engine." TheBackSeatPilot.com.

Time Magazine. "Death in Grindstone Canyon." July 20, 1953.

U.S. Department of the Interior National Park Service, National Register of Historic Places Registration Form. https://ohp.parks.ca.gov/pages/1067/files/CA_Glenn%20County_Willows-Glenn%20County%20Airport_DRAFT.pdf

Wilkinson, Stephan. "Firebombers! Flying on the Edge to Fight Fires." HistoryNet, January 24, 2020.

Wyckoff, Bob. "Aerial Firefighting Began with Military Aircraft." *The Union*, July 23, 2007.

ACKNOWLEDGMENTS

My family—I owe them everything.

And Joe Layden, whose mentorship and expertise put this book in your hands.